THE
VICTORIAN
SCIENTIST

THE VICTORIAN SCIENTIST

The Growth of a Profession

JACK MEADOWS

THE BRITISH LIBRARY

First published 2004 by
The British Library
96 Euston Road
London NW1 2DB

British Library Cataloguing-in-Publication Data
A catalogue record for this book is available from The British Library

ISBN 0 7123 0894 6

Designed by Bob Elliott
Typeset by Hope Services (Abingdon) Ltd
Printed in England by Cromwell Press
Trowbridge, Wiltshire

Contents

Acknowledgements

I am greatly indebted to Professor Bill Brock for reading the manuscript and making a variety of valuable suggestions. I would also like to thank the Information Science department at Loughborough University and the libraries of both Loughborough and Leicester Universities for allowing me access to their facilities. I am grateful to Georgina Difford and Tony Warshaw at the British Library for all their work on the production of this book, and to Kathy Houghton for chasing up the illustrations.

Introduction

HISTORY textbooks will tell you that the origins of modern science are to be found in the seventeenth century. In terms of the basic ideas, this is quite true. But, for the building of the modern scientific community, the nineteenth century is the key period. That was when science became important in education and in industry; that was when scientists became professionals with a significant role in society. The interesting thing is how few people were involved in bringing about this change. Francis Galton estimated that there were only some three hundred scientists of any eminence in Britain in the 1870s. The number of eminent scientists whose names were recognised by Victorian society at large was appreciably less – no more than forty or fifty. So who were this small band of brothers (there were no female members) who brought science to the centre of the stage, and how did they do it?

The lives of most leading British scientists of the period have been recorded in some detail by themselves, their contemporaries and subsequent writers. Though such writings must always be taken with a pinch of salt, how these scientists saw the world, and how the world treated them, can be fairly well determined. This book is a kind of collective biography that records the fortunes of eminent Victorian scientists – for the most part in their own words. It is a record of the development not of the science, but of the scientist. Yet following scientists' careers in parallel does provide some idea not only of their own enthusiasms and problems, but also of how science itself evolved in the nineteenth century. In addition, it shows scientists as human beings interacting both with each other and with the rest of society.

Looking back, the pace of change in the scientific community seems greatest in the period from the 1850s to the late 1870s. So this book will concentrate particularly on scientists who were active during these years. At the same time, it will be necessary to look at some of their predecessors, who taught them their trade and created the scientific milieu in which they grew up. It will also be necessary to look at comments by some of their successors, whom they, in turn, influenced. Of course, such time divisions are arbitrary: after all, some scientists were active for a good portion of the century. George Airy, for example, was appointed Astronomer Royal (the senior post in British astronomy) in 1835 and only retired in 1881. But this overall picture is supported by contemporary comments. Beatrice Webb, the social reformer, was born in the 1850s. She later commented on, 'the belief in science and the scientific method, which was certainly the most salient, as it was the most original, element of the mid-Victorian Time-Spirit'.[1] The keyword here is 'mid-Victorian', remembering that Queen Victoria came to the throne in the 1830s and died at the beginning of the twentieth century.

During the Victorian period, there were great changes in the material environment, especially in terms of communication. A reliable and rapid postal system, an extended railway network, the telegraph and, later on, the telephone all linked people together in a way that had not previously been possible. Scientists were keen users of these new means of communication, and, indeed, increasingly contributed to their development. Much of this book is necessarily concerned with the lives of scientists in Britain. But the nineteenth century saw a rapid growth in interaction between British scientists and their colleagues overseas, stimulated and aided by better communications. So it is also necessary to look at British scientists in terms of the international scene. There is one point to notice. Though I refer throughout to 'scientists', the name was only coined in the 1830s (by Whewell) and many scientists were initially opposed to it. They preferred, until the latter years of the century, to be known as 'men of science'. Besides, there were already words in existence to describe their specific interests. Such names as 'chemist' and 'geologist' were in circulation when the nineteenth century began. Indeed, 'scientist' was modelled on these existing words. Physics came a little later on the scene as an integrated subject, so Whewell similarly coined the word 'physicist' to describe its practitioners. For the most part, they disliked it. Faraday claimed that he could not even pronounce the word.

Scientists in the nineteenth century, like other professional people, had to worry about making a living. Their problem was greater because, until the last years of the century, science was not really regarded as a profession: there was no standard method of acquiring a job and a salary, as there was in traditional professions such as medicine or law. Consequently, money questions crop up quite often in accounts of the lives of Victorian scientists. Two points need to be made. The first is that, though the value of the pound varied appreciably throughout the nineteenth century, there was no major inflationary trend overall. The following figures give a general idea of the value, in present-day terms, of £100 in Victorian times: 1830 – £5,400; 1850 – £6,400; 1870 – £5,000; 1890 – £6,500.[2] This means that a given amount of money was of roughly comparable value across the entire Victorian period. The other point is that the money system then in use is now fading from human memory. Here is a brief reminder. The basic unit was, as it still is, the pound (£). This was divided into 20 shillings (s), and each shilling into 12 pence (d). Professional people were often paid in guineas (one pound and one shilling). William Crookes complained to Edward Frankland about the cost of transcribing and publishing some of the latter's lectures. It would require, he said, £43/7/- (that is 43 pounds, 7 shillings, and no pence). Since this comes out at some £2,500 in present currency, we can understand his anguish.

Both Crookes and Frankland were members of the small group of eminent nineteenth-century scientists, but they are hardly household names today. The one name that everybody does remember is Charles Darwin. I have therefore used Darwin as my recurring example in this book. Exploring to what extent he was a typical nineteenth-century scientist, and to what extent he differed from his fellows, helps build up a picture of the group to which he belonged. Because memories of most of these scientists have faded, I include, immediately after this *Introduction*, a set of brief biographies of the characters who crop up most frequently throughout the book. They are divided into three groups, depending on when they were born – pre-1800, 1800–1830, and 1830–1860. (Scientists born after 1860 hardly had time to become eminent before the end of the nineteenth century.) People not on this list are described briefly as they appear in the book. Some of the scientists listed here received knighthoods. (Thus Crookes and Frankland ultimately became Sir William Crookes and Sir Edward Frankland.) A few were elevated further. Lister, for example, became Baron Lister. These honorifics are omitted here. The exception is William Thomson, who later became Baron Kelvin. There were so many Thom(p)sons in nineteenth-century science that it has become customary to refer to him simply as 'Kelvin'. Rayleigh, who also appears in the list, is a rather different case, since he inherited his title.

THE MAIN ACTORS

BORN BEFORE 1800

Charles Babbage (1792–1871) Born in Devon. Studied mathematics at Cambridge University, where he helped introduce new mathematical ideas from the Continent. Professor of mathematics at Cambridge for some years. Babbage spent much of his life trying to construct a mechanical computer, using both his own money and funding from the Government. Though he obtained some valuable results, his ideas never came fully to fruition.

David Brewster (1781–1868) Born in the Scottish borders. Studied at Edinburgh University. Later became Principal, first of St. Andrews University, then of Edinburgh University. His main scientific interest was optics, but he was also a noted journal editor and populariser of science.

William Buckland (1784–1856) Born in Devon. Educated in Oxford where he also became an ordained clergyman. Held posts in mineralogy and geology at Oxford, but ended his career as Dean of Westminster. He was a major authority on fossils and invoked catastrophes as an explanation of geological change.

John Dalton (1766–1844) Born in Cumberland. A Quaker who was mainly self-taught in science. Spent much of his life as a teacher or private tutor in Manchester. His most important work was concerned with the properties of gases and atomic theory.

Humphry Davy (1778–1829) Born in Cornwall. Initially apprenticed to a local surgeon, but became interested in chemistry. Appointed to the newly-founded Royal Institution in London, which he soon made into a centre for popularising science. Davy discovered a number of new elements, but was best-remembered in the nineteenth century for his invention of the miner's safety lamp.

Michael Faraday (1791–1867) Born in Surrey. Member of a small Protestant sect called the Sandemanians. Self-taught in science, becoming Davy's assistant at the Royal Institution in his early twenties. Later succeeded to Davy's post as professor there. Faraday carried out important work in chemistry, but his most significant research was concerned with electricity and magnetism. A major figure in the popularisation of science.

John Herschel (1792–1871) Born near London. The only son of the famous astronomer, William Herschel. Trained in mathematics at Cambridge, where he helped introduce new mathematical ideas from the Continent. Herschel subsequently carried out extensive astronomical observations in the Southern hemisphere, and became a pioneer contributor to the science of photography.

Charles Lyell (1797–1875) Born in Scotland. Educated at Oxford University. Supported the idea of gradual, rather than catastrophic change in geology. His book, *Principles of geology*, was the most influential geology text for much of the nineteenth century. It was important for the development of Darwin's ideas on evolution.

Roderick Murchison (1792–1871) Born in Scotland. He became an Army officer, developing an interest in geology after he was discharged from the Army. He did pioneering work in distinguishing new geological strata. Became Director-General of the Geological Survey.

William Whewell (1794–1866) Born in Lancaster. Educated at Cambridge University, becoming an ordained clergyman. Appointed Professor of Mineralogy at Cambridge, and subsequently Professor of Moral Theology there. He worked on tidal theory, but is best remembered for his writings on the history and philosophy of science.

BORN 1800–1830

George Airy (1801–1892) Born in Northumberland. Read mathematics at Cambridge University. Professor of Mathematics there, then Astronomer Royal at Greenwich Observatory for nearly fifty years. He made major changes in the work at Greenwich, and developed new instrumentation.

George Boole (1815–1864) Born in Lincoln. Largely self-taught in mathematics. Later became Professor of Mathematics at Cork University in Ireland. Best known for his influential work on mathematical logic.

Arthur Cayley (1821–1895) Born in Surrey. Read mathematics at Cambridge. Was in legal practice for a number of years, but continued mathematical research. Became professor at Cambridge in 1863. Published nearly a thousand papers during his lifetime.

Charles Darwin (1809–1882) Born in Shropshire. Educated at Edinburgh and Cambridge Universities. Joined *HMS Beagle* as naturalist in 1831. Subsequently became leading authority on natural history and geology. Published his major work, *Origin of species*, in 1859. Owing to ill-health, Darwin was less involved in scientific meetings in the latter years of his life.

Edward Frankland (1825–1899) Born in Lancashire. Studied chemistry in London and Germany. Professor in Manchester and London, ending his career at South Kensington. Developed ideas on chemical combination. On the practical side, he was an expert on sanitation problems.

Francis Galton (1822–1911) Born in Birmingham. A half-cousin of Darwin. Studied medicine at Birmingham and London, then mathematics at Cambridge. Spent some time exploring in Africa. He had a wide range of scientific interests, including meteorology, heredity, statistics, and fingerprints.

Joseph Hooker (1817–1911) Born in Suffolk. Educated at Glasgow University. A botanist and friend of Darwin, Hooker went on various overseas expeditions to collect material. He succeeded his father as Director of the Royal Botanic Gardens at Kew in 1865.

Thomas Huxley (1825–1895) Born in Middlesex. Studied medicine in London. Then joined *HMS Rattlesnake* on its voyage to the South Seas. Became Professor of Natural History at South Kensington, where he played a major role in zoological classification. Huxley was an energetic advocate of Darwin's ideas, and a leading publicist for science.

James Joule (1818–1889) Born near Manchester. Tutored at home by Dalton. Like Dalton, he became an important figure in science in Manchester. Famous for his studies of heat and energy.

Kelvin *See* William Thomson

Joseph Lister (1827–1912) Born in Essex. Read medicine in London. He subsequently held medical posts in both Scotland and London. His studies led him to introduce the practice of antiseptic surgery.

Richard Owen (1804–1892) Born in Lancaster. Studied medicine in Edinburgh and London. Became Curator of the museum at the Royal College of Surgeons. Then took charge of the British Museum of Natural History. Leading expert on comparative anatomy. Opposed Darwinian ideas.

Lyon Playfair (later **Baron Playfair**) (**1819–1898**) Born in India. Studied chemistry at universities in Scotland, London and Germany. Worked for the Science & Art Department at South Kensington. Then became Professor of Chemistry at Edinburgh. Spent the latter part of his career as a Liberal M.P.

Herbert Spencer (**1820–1903**) Born in Derby. Self-taught philosopher. Became one of the founders of sociology. Strong advocate of evolutionary ideas.

George Stokes (**1819–1903**) Born in Ireland. Read mathematics at Cambridge University, and became Professor of Mathematics there. He had wide-ranging interests in theoretical physics, especially the motion of fluids. Was a Conservative M.P. for a few years.

James Sylvester (**1814–1897**) Born in London. Studied mathematics at Cambridge University, but was unable to take his degree for many years because of his Jewish faith. Held posts as a professor of mathematics both in London and in the United States.

William Thomson (later **Baron Kelvin**) (**1824–1907**) Born in Ireland. Studied at Glasgow University, then mathematics at Cambridge University. Became Professor of Natural Philosophy at Glasgow University. Wide-ranging interests in physics and the design of instrumentation. He did pioneering work in telegraphy and the construction of the first transatlantic cable.

John Tyndall (**1820–1893**) Born in Ireland. Studied physics in Germany. Appointed Professor of Natural Philosophy at the Royal Institution. Noted for his work on heat radiation, and as a populariser of science.

Alfred Wallace (**1823–1913**) Born in Wales. Self-taught in science. Travelled widely as a naturalist. Arrived at the basic ideas of evolution before Darwin published his *Origin of species*. Pioneered work on the geographical distribution of animals.

BORN 1830–1860

Henry Armstrong (**1848–1937**) Born in London. Studied chemistry in London and Germany. Became Professor of Chemistry at the Central Technical College. Particularly noted for his work on education.

William Clifford (1845–1879) Born in Devon. Educated in London, then read mathematics at Cambridge University. Became professor at University College London. Though he died young, he was an important populariser of science.

William Crookes (1832–1919) Born in London. Studied chemistry in London. He worked as a freelance consultant and editor for much of his life. Particularly noted for studies involving vacuum tubes.

Norman Lockyer (1836–1920) Born in Warwickshire. Self-taught astronomer, and a pioneer of astronomical spectroscopy. Became a professor at South Kensington. He was founder-editor of the journal *Nature*.

Oliver Lodge (1851–1940) Born in Staffordshire. Studied mathematics and physics in London. Became professor at Liverpool University and then Principal of Birmingham University. Noted especially for his work on radio waves.

John Lubbock (later Baron Avebury) (1834–1913) Born in London. Self-taught in science. Pioneering work in anthropology and animal behaviour. Liberal M.P. for many years.

James Clerk Maxwell (1831–1879) Born in Edinburgh. Studied at Edinburgh University, then mathematics at Cambridge University. He was professor successively in Aberdeen, London and Cambridge. Despite his early death, he did very significant work in physics, especially on electricity and magnetism.

William Ramsay (1852–1916) Born in Glasgow. Studied chemistry in Germany. Professor, then Principal, of University College Bristol. Was subsequently Professor of Chemistry at University College London. Noted for his work on the noble gases.

Rayleigh (John Strutt, Lord Rayleigh) (1842–1919) Born in Essex. Read mathematics at Cambridge University. Succeeded Maxwell as Professor of Physics there. Subsequently, followed Tyndall as a professor at the Royal Institution. Much of his research was carried out privately at home. Wide-ranging interests in theoretical and experimental physics.

Henry Roscoe (1833–1915) Born in London. Studied chemistry in London and Germany. Became Professor of Chemistry at Manchester.

Was a Liberal M.P. for a number of years. He was particularly noted for his work on education in science.

Joseph John Thomson (1856–1940) Born in Manchester. Educated at Manchester and Cambridge, where he read mathematics. Succeeded Rayleigh as Professor of Physics at Cambridge. Carried out pioneering work on the nature of the atom. Widely known from his initials as 'J. J.'

Peter Tait (1831–1901) Born in Scotland. Educated at Edinburgh and Cambridge, where he read mathematics. Became Professor of Natural Philosophy at Edinburgh University. He worked especially on the motion of fluids and heat.

The biologist Charles Darwin looking particularly satisfied, perhaps because he had just published *The Descent of Man*. From *Vanity Fair* (30 September 1871). Newspaper Library, Colindale

School and home

SCHOOLDAYS were not the happiest days of Charles Darwin's life. He particularly disliked the emphasis on the classics. He later commented ruefully that: 'Nothing could have been worse for the development of my mind than Dr. Butler's school, as it was strictly classical'.[1] Darwin's school at Shrewsbury was a public school, meaning that it catered for the sons of the wealthier members of the population. Under Samuel Butler, it had grown greatly both in size and in its academic reputation. Indeed, Darwin was rather unfortunate in having a headmaster who actually spearheaded the extension of classics (that is Latin and Greek) teaching in the early nineteenth century. For Butler, a school with a good reputation was one with a good reputation for teaching classics (though this might include some mathematics, mainly in the form of Euclid's geometry). His grandson – also Samuel Butler, and later an anti-Darwinian writer – attended Shrewsbury School, and described the curriculum. It appears that subjects other than the classics were only taught on what were really supposed to be school holidays.

Darwin's interests as a boy ran rather to natural history and chemistry. Natural history, in the early nineteenth century, was borderline as an acceptable hobby for a boy. Chemistry was something else again. Darwin's father allowed Charles and his brother to do chemical experiments in a garden shed. Indeed, he allowed them to spend £50 on chemical equipment: the equivalent of the annual wage for many lower-paid workers in those days. Unfortunately, news of this activity leaked out:

The fact that we worked at chemistry somehow got known at school, and as it was an unprecedented fact, I was nicknamed 'Gas'. I was also once publicly rebuked by the headmaster, Dr. Butler, for thus wasting my time on such useless subjects.[2]

Darwin's experiences raise three obvious queries about the schooldays of Victorian scientists. Did the usual school curriculum turn them off?

Were they already interested in science when they were young? How important was it to have supportive (and preferably wealthy) parents?

The joys of school

Only a limited number of eminent Victorian scientists went to public schools. Those who did generally agreed with Darwin's opinion. For example, John Lubbock went to Eton when he was aged eleven. He found the classical curriculum boring, and was delighted when his father agreed, after four years, that he could leave school and start work in the family firm. A greater number of scientists attended grammar schools, where the curriculum, like that of public schools, was for many years dominated by the classics. (As the word 'grammar' indicates, they were actually founded to teach Latin and some Greek.) Edward Frankland went to Lancaster Free Grammar School (actually the most expensive school in Lancaster, despite its name) as Queen Victoria came to the throne. He commented in later years on how much he hated the Latin taught there. Alfred Wallace attended the grammar school in Hertford at much the same time, and subsequently remarked on the boredom induced by rote learning of Latin. A few scientists, later to be eminent, enjoyed the activity. George Airy went to Colchester Grammar School, and recalled his facility at rote learning with pride.

It was the custom for each boy once a week to repeat a number of lines of Latin or Greek poetry, the number depending very much on his own choice. I determined on repeating 100 every week, and I never once fell below that number and was sometimes much above it. It was no distress to me, and great enjoyment. At Michaelmas 1816 I repeated 2394 lines, probably without missing a word.[3]

This was not the usual reaction. More scientists were likely to sympathise with Richard Carrington – later an eminent astronomer – in the way he remembered such teaching:

I consider that all attempts to make me learn . . . Latin poetry were gross mistakes; I was never benefited in the least. Reasoning was my forte, and I could never do anything by rote.[4]

Airy was to become a great believer in the importance of routine work in science, but even he despised the methods adopted for teaching Latin in schools. William Whewell attended the same school as Frankland, though several years earlier. He actually enjoyed the classics, but this may simply have reflected his interest in almost anything. The

Reverend Sydney Smith remarked of Whewell in later life that 'Science is his forte, and omniscience his foible'. Humphry Davy, who was at Truro Grammar School for a while towards the end of the eighteenth century, tersely summarised his reaction to the curriculum. He wrote to his mother: 'Learning is a true pleasure; how unfortunate then it is that in most schools it is made a pain'.[5]

It was not uncommon to experience a range of schools. Francis Galton, another polymath like Whewell, passed successively through home tuition, a dame's school, a French school, a private school and then King Edward's School in Birmingham. His experience of the latter was similar to that of his half-cousin, Charles Darwin, in Shrewsbury. He later explained that:

[I] craved for what was denied, namely, an abundance of good English reading, well-taught mathematics, and solid science. Grammar and the dry rudiments of Latin and Greek were abhorrent to me, for there seemed so little sense in them.[6]

Many future nineteenth-century scientists received part, or all, of their early education either at home, or at a private school. The reasons were diverse. Sometimes, a father might take his family with him when he went abroad, in which case recourse to private tutors was usual. So George Bentham, later a leading botanist, received private tutoring at various times in Russia, France and England. Another common reason was poor health. James Joule suffered from a spinal defect, and was tutored at home. Lord Rayleigh spent short periods at both Eton and Harrow, but was mainly schooled privately because of health problems. To survive a public school needed robust health in those days. John Herschel was sent to Eton at the age of eight. He was removed and sent to a private school when his mother saw him being knocked about by one of the bigger boys. Sometimes it was simply a question of parental wishes. Lord Rosse, who subsequently became a famous astronomer, was perfectly healthy, but his parents wished to tutor him at home themselves. In all these cases, there was ample money available to educate the boy privately. At the other end of the scale, people with a limited supply of money were also likely to use private or home tuition. George Boole was the son of a cobbler. His father tutored him at home up to the age of ten, and then persuaded a local bookseller to act as a tutor.

Private or home tuition tended to be more diversified in terms of curriculum than that provided by public and grammar schools.

The physicist Lord Rayleigh; who was actually a good deal less ferocious than this cartoon suggests. From *Vanity Fair* (21 December 1899).
Newspaper Library, Colindale

Although classics still figured in the teaching, it was usually accompanied by other subjects. Charles Pritchard, who later became Professor of Astronomy at Oxford University, attended a private school where the teacher provided tuition in practical activities, such as the use of a theodolite for surveying. Joule was taught by John Dalton, one of the most famous scientists in the country in the early nineteenth century. (Despite his fame, Dalton made his living as a tutor for much of his life.) Dalton was a Quaker, and the education of non-conformists was usually less concerned with the classical side of the curriculum. Part of the reason for the emphasis on classics in England was to provide for entry to Oxford or Cambridge Universities, where non-conformists were unable to take degrees until the latter part of the nineteenth century. Perhaps this lesser emphasis on the classics made the subject more acceptable to the pupils. Boole took great pride as a schoolboy in the Latin funeral oration he wrote when a pet rabbit died. Lord Kelvin was tutored at home by his father. He later explained where he thought the classics should fit into the curriculum.

A boy should have learned by the age of twelve to write his own language with accuracy and some elegance; he should have a reading knowledge of French, should be able to translate Latin and easy Greek authors, and should have some acquaintance with German.[7]

Kelvin's father was a professor of mathematics, first in Belfast and then in Glasgow. Scotland and, to a lesser extent, Ireland had a reputation for providing a broader education than England. David Gill, who later devoted his life to astronomy, went to Dollar Academy in Scotland, which had a headmaster interested in science. But differences from England were not always so evident in curricula at the school level. Lyon Playfair commented adversely on the large amount of Latin taught at his grammar school. David Forbes, subsequently one of a group of eminent Scottish geologists, observed that the education in Latin and Greek he received at school was a positive deterrent to a scientific outlook. Another Scottish scientist summarised his schooling as: 'Omission of almost everything useful and good, except being taught to read. Latin! Latin! Latin!'.[8] James Clerk Maxwell and Peter Tait were friends at the Edinburgh Academy, where the emphasis was firmly on classics and mathematics. Maxwell found the Latin component unappealing – to the extent that he was nicknamed 'Dafty' by his schoolmates. As in England, private schools were sometimes better in this respect. John Tyndall, though a Protestant, attended a Roman

Catholic school in Ireland. The headmaster, an excellent mathematics teacher, provided practical tuition in surveying.

It required a Parliamentary Act in 1840 before grammar schools could legally diversify their curriculum. They did so slowly. In the 1860s, Oliver Lodge went to Newport Grammar School, where he disliked the continuing emphasis on Latin. William Perkin, the chemist, was more fortunate. He attended the City of London School in the 1840s. This was an early example of a new type of school that appeared in the nineteenth century. Such schools took a broader view of the curriculum than the existing public schools. By the time Perkin arrived there, the City of London School had on its staff a teacher who had been trained at the Royal College of Chemistry. At the time, this was the only college in the country specialising in chemical education. In consequence, unlike Darwin, Perkin was encouraged in his chemical interests by the school. He later sent his sons to the same school and they, in turn, also became chemists. Henry Roscoe attended a similar school in Liverpool, also in the 1840s. His formative experiences there explain why chemistry was popularly known as 'stinks'.

Each boy was provided with a glass containing powdered sulphide of iron and with a second one containing dilute sulphuric acid. 'When I give the command', said [the teacher], 'each boy will pour the acid on the sulphide, and you must then all run away as fast as your legs can carry you'. No sooner said than done! The result was such a fearful stench that each boy will carry down the recollection of that moment to his grave.[9]

In the 1860s, the pace of educational change in schools began to increase. The most obvious sign of this was the succession of Royal Commissions that were set up to investigate education. The Newcastle Commission reported in 1861 on elementary education, followed by the Clarendon Commission in 1864 on public schools, and the Taunton Commission in 1868 on grammar schools. But this activity was too late for our group of eminent scientists. It is difficult to be recognised as eminent in science at an age less than the mid-thirties. This means that all our eminent scientists had received their schooling by end of the 1860s, just as this era of educational discussion was beginning. J. J. Thomson went to school during that decade, and subsequently recalled:

It was a school which was not affected by the new views on education which were just then coming in. We were taught Latin from the Eton Latin Grammar, where the rules for syntax and grammar were in Latin, and we learnt these by heart before we knew what they meant.[10]

Darwin's headmaster called chemistry 'useless'. In the world of nineteenth-century public schools, which were, after all, actually expensive private schools, the use of words must be interpreted with care. The public school ethos was centred on the idea of a 'liberal education'. This essentially meant an education which developed the character, which trained the mind to be flexible, and which provided the future leaders of the nation with a common cultural background. For these purposes, the classics were believed to be uniquely appropriate. Mathematics was given a subsidiary place in the curriculum, not least because it was held to be important by Cambridge University. But most of the older schools believed its educational value to be decidedly inferior. At Eton in the mid-nineteenth century, the mathematics teachers employed there were not allowed to wear gowns, or to become housemasters (unlike their colleagues teaching classics), and they received lower salaries.

So when Dr. Butler said chemistry was useless, he meant that it was not educative in the sense that the classics were. He would have agreed that chemistry was useful in practical terms, but would have added that, should any poor benighted souls need a knowledge of chemistry in later life, then the flexible mind produced by a classical education would readily allow them to acquire it. Whatever the way ahead, classics provided the essential background. As a Cambridge don of the time explained to his students, 'the study of Greek literature . . . not only elevates above the common herd, but leads not infrequently to positions of considerable emolument'.[11] Not everybody agreed with this argument. In the early 1840s, William Thackeray, the novelist, pointed out to students at an Irish agricultural college:

You are not fagged and flogged into Latin and Greek at the cost of two hundred pounds a year. Let these be the privileges of your youthful betters; meanwhile content yourselves with thinking that you *are* preparing for a profession while they are *not*.[12]

Science education also had another black mark against it – it smelt of radicalism. For example, the pressure for reform exerted by the Chartists was a cause of much worry to the better-off. One item in the Chartist programme was the creation of schools that included science in their curriculum. Materialistic science was seen as having connections with the revolutionaries in France – sometimes not without reason. Early in his career, Davy worked as an assistant to Thomas Beddoes, who was interested in the medical uses of gases.

Beddoes had retired from his post at Oxford University not long before because his sympathies were with the revolutionaries – not a popular attitude at Oxford. Davy had to make his way carefully as a result of the suspicions his connection with Beddoes raised.

This survey of opinions suggests that Darwin's reaction to his schooling is equally reflected in the reactions of many of his contemporaries. The emphasis on the classics generally ran counter to their main interests. From this viewpoint, the higher the school in terms of prestige, the less likely it was to assist a pupil's scientific inclinations. It is interesting to see how Charles Darwin dealt with the education of his own sons. Like a typical country gentleman, he sent his eldest son, William, to a public school – in this case, Rugby – where he settled in happily. But Darwin noted that the steady diet of classics at the school seemed to be contracting his son's range of interests and powers of observation. (To no one's surprise, William later became a banker.) So he decided to send his other sons to a school where the studies were more diversified. He chose Clapham Grammar School, then in the charge of Charles Pritchard, a contemporary of Darwin's at Cambridge. Pritchard, as befitted someone interested in astronomy, provided a curriculum that included mathematics and science. (Darwin was not the only eminent scientist who wanted this type of curriculum for his offspring. John Herschel's sons also went to Pritchard's school.) Darwin reverted to type for his daughters, who had a governess at home.

Getting involved

Darwin showed an early interest in science. Was this true also of his fellow scientists? Several tried their hand at experiments from early on. Frankland dabbled in chemistry from the age of seven. William Crookes similarly carried out experiments from an early age. His family allowed him to convert a cupboard in the house into a chemical laboratory, but then objected strongly to the smells he produced. Spectacular chemical experiments were especially popular. When Crookes was fourteen, gun-cotton was discovered. He proceeded to make some, and almost blew himself and his friends up. Robert Ball, the astronomer, seriously injured himself as a boy when he tried to make fireworks. Rayleigh horrified his parents by buying gunpowder for his experiments. Safety was usually learned the hard way. Years later, Lord Kelvin commented: 'Lord Rayleigh has told us that he burnt his fingers with phosphorus when he was only twelve years old. I burnt mine in the same way at eighty-two'.[13]

Alongside such experimenting, a number of future scientists enjoyed collecting. Darwin was omnivorous in this respect. He was prepared to collect anything, and remarked later that it was a passion shared by naturalists and misers. Others were more specialist. Lubbock, for example, was keen on collecting insects from early on in his life. So, too, was Charles Lyell. He later recalled his activities:

Collecting insects was just the sort of desultory occupation which suited me . . . as it gave sufficient employment to my mind and body, was full of variety, and to see a store continually increasing, gratified [my] 'accumulative propensity' . . . I had no companion to share this hobby with me, no-one to encourage me in following it up, yet my love for it continued always to increase. . . . I was very ill supplied with instruments, both for taking and preserving my prey, and many insects of real value to the entomologist, as I afterwards discovered, were mutilated by being knocked down by my hats . . . Instead of sympathy, I received from almost everyone else beyond my home, either ridicule, or hints that the pursuits of other boys were more manly. . . . The disrepute in which my hobby was held had a considerable effect on my character, for I was very sensitive of the good opinion of others, and therefore followed it up almost by stealth.[14]

A mechanical bent was quite common. Airy made mechanical models. James Dewar, later a leading experimental physicist, made violins with the aid of a local joiner. Herschel received instruction in the use of tools from his father's workmen, and applied them in an attempt to demolish part of his home (his parents caught him in time). Of all the major scientists, perhaps the two most experimentally inquisitive as children were James Clerk Maxwell and Charles Babbage. Maxwell's interests covered all the world around him. His mother wrote of him when he was two years old: 'He has great work with doors, locks, keys, etc and "Show me how it does" is never out of his mouth. He also investigates the hidden course of streams and bell-wires'.[15] Babbage was equally investigative from an early age:

My invariable question on receiving any new toy, was 'Mamma, what is inside of it?' Until this information was obtained those around me had no repose, and the toy itself, I have been told, was generally broken open if the answer did not satisfy my own little ideas of the 'fitness of things'.[16]

Babbage pushed his experimental investigations further than most: he decided to see whether it was possible to raise the devil. He records the result in his autobiography:

A signed portrait of the geologist Charles Lyell by the Royal Academician George
Richmond. From Mrs. Lyell, *Life, letters and journals of Sir Charles Lyell*
(John Murray, 1881).
The British Library, 10854.g.10

As I only desired an interview with the gentleman in black simply to convince my senses of his existence, I declined adopting the legal forms of a bond, and preferred one more resembling that of leaving a visiting card . . . Accordingly, having selected a promising locality, I went one evening towards dusk up into a deserted garret. Having closed the door, and I believe opened the window, I proceeded to cut my finger and draw a circle on the floor with the blood which flowed from the incision. I then placed myself in the centre of the circle, and either said or read the Lord's Prayer backwards. . . . I then stood still in the centre . . . looking with intense anxiety in all directions . . . Fortunately for myself, and for the reader also, if he is interested in this narrative, no owl or black cat or unlucky raven came into the room.[17]

One of the popular activities in mid-Victorian times was visiting a huge variety of 'entertainments'. Some of these had a scientific flavour, including everything from exhibitions of strange people or animals to demonstrations of new inventions (both categories often involving some deceit). Among these entertainments were ones which involved people with some training in science. A famous example was 'Pepper's ghost'. Henry Pepper lectured on science at the Royal Polytechnic in Regent's Street and helped develop an optical device for projecting what appeared to be a ghostly figure onto a stage. He reported on this and other activities in a series of popular science books. Future scientists were apt to be fascinated by such entertainments. Crookes, for example, who lived close to the Polytechnic, was inspired by Pepper as a boy.

Sport became an equally popular entertainment during the nineteenth century. Most boys played games at school. Wallace, for example, remembered playing cricket and baseball: football became popular later in the century. Baseball was played along with cricket in the early years of the nineteenth century, but cricket came to dominate in later years. J. J. Thomson, who was educated in the mid-century, remembered playing both cricket and football – the latter being the rugger variety:

Football then was a rather gruesome business; 'hacking' was allowed, that is, the forwards in the scrum just kicked at anything in front of them, whether it was the ball or the shins of the opponents; the result was that after a game one's shins were always bruised and often bleeding.[18]

Few leading scientists continued team games in later life, but some of the less organised activities, picked up in boyhood, were continued. Boys from wealthier families – Darwin is an example – often enjoyed

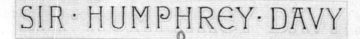

SIR · HUMPHREY · DAVY

THE INVENTOR OF THE SAFETY LAMP

SIR HUMPHREY DAVY.
born 1778. died 1829.

T. NELSON AND SONS.
LONDON, EDINBURGH AND NEW YORK.

This portrait of the chemist Humphry Davy celebrates his invention of the safety lamp for use in mines. The nineteenth-century writer, like many since, misspells Davy's first name as 'Humphrey'.
The British Library, 10602.aaaa.14

shooting and riding (though the latter activity was, in any case, necessary in the early years of the century simply to get about). Fishing was cheaper, and so more widely popular. Herbert Spencer recalled making his own tackle because he could not afford to buy any with his pocket money. But the greatest devotee was certainly Humphry Davy. In later years, he published a book called *Salmonia, or days of fly-fishing*, and an American visitor noted:

I found that the first chemist of his time was a professed angler; and that he thinks, if he were obliged to renounce fishing or philosophy, that he would find the struggle of his choice pretty severe.[19]

The overall impression is that our group of scientists, as boys, were frequently involved in activities with a scientific slant. Often this can be related to their later interests, but not always. J. J. Thomson, though later a famous physicist, was mainly interested in botany as a boy, an interest which he kept up throughout his life. Henry Armstrong, later a chemist, collected enthusiastically with a friend, when young, and had a keen interest in geology. His friend's collection of butterflies and moths was continued and subsequently became part of the collection at the British Museum (Natural History). Equally, people later famous in biology might have an early interest in physical science. The young Huxley was especially interested in mechanical things, and thought of becoming a mechanical engineer. So did his contemporary in biology, William Carpenter, whose nickname at school of 'Archimedes' suggests well enough where his interests lay. Of course, not all boys had sufficient leisure to pursue such hobbies. Michael Faraday, for example, coming from a poor background, had no opportunity to think about science until he was already working.

Scientific hobbies became more respectable as the century progressed. By the end of the century, books for boys discussed scientific hobbies as well as sporting activities; and attempts were even made to cater for boys from different backgrounds. As one editor explained, alternatives were provided so that boys of all ages and conditions could try their hands, whether they had, 'leisure and ample opportunities, or [were] already closely engaged in the sterner duties of bread-winning; boys to whom a considerable preliminary outlay may be of trifling moment, and others who rarely have a shilling to spare'.[20] Though, in the early years of the century, Lyell had to disguise his interest in collecting, by J. J. Thomson's time, it had become an acceptable activity:

Natural history was pressed into the service of boys when I and some of my friends tried to raise money to buy wickets, bats and balls for a small cricket club we wanted to start, by having an exhibition of butterflies, moths, birds' eggs and wild flowers which our relatives were expected to attend, and pay for admission.[21]

What to read

Along with experimenting and collecting, many of the boys were encouraged in their scientific bent by their reading. Popular science books were few and far between in the early years of the nineteenth century. Seekers after scientific information had to make do with a rather assorted bag. One book frequently mentioned as providing a stimulus was *Sandford and Merton*. (The author, Thomas Day, was a radical thinker, a contemporary of Erasmus Darwin, and a disciple of Rousseau in terms of educational theory.) Spencer read the book in the 1820s, Frankland in the 1830s and Oliver Lodge as late as the 1850s. By that time the book had been in print for seventy years. Merton, in the story, is the son of a rich merchant, while Sandford is the son of a local farmer. The latter gradually convinces the pampered Merton that it is important to gain a practical understanding of the world around them. Frankland later explained the value of the book:

The chief attraction for me in this book consisted in the exposition of natural laws, in language which I could understand. It did not then seem to me anomalous that the author should consider it likely that most of his readers would be cast upon desert islands and be compelled to do and make everything for themselves . . . It certainly gave me the first impetus to observation as distinguished from hearsay.[22]

The first science book that attracted the young Darwin's attention was *Wonders of the world* by the Revd. C. C. Clarke. This book was one of a whole phalanx of popular and widely read works on nature produced by parson-naturalists throughout the nineteenth century. In 1858, the year before Darwin's *Origin of species* appeared, the Revd. J. G. Wood's *Common objects of the countryside* was published. It probably holds the publishing record for this kind of book, for it is said to have sold 100,000 copies in a week. Some introductory chemical texts were also available early in the century: Mrs. Marcet's *Conversations on chemistry* was particularly popular. It was from this book that Faraday learnt about electrochemistry, which was to become one of his special research interests. There were also a few introductory texts on physics, one of the popular ones again having a female author – Mrs. Bryan's *Lectures on natural philosophy*. Several authors took an experimental approach. Darwin read Samuel Parkes' *Chemical catechism*, which fell into this category. Not all the writers of such books were experts. Spencer recalled:

> Little by little I became much interested in chemistry at large, and read with interest a small book by an itinerant lecturer named Murray, who at that time occasionally came to Derby – a very incompetent man, but one who served to make familiar the simpler truths.[23]

Because the number of lower-level books on science was limited in the early years of the nineteenth century, some of the budding scientists had recourse to more advanced texts. Davy read Lavoisier's *Elementary treatise on chemistry* (in the original French). Huxley read Hutton's *Geology*, which, like Lavoisier's book, is still regarded as a classic text in its field. Darwin used Brande's *Manual of chemistry*. (William Brande, a leading chemist of his day, had written the book primarily as a text for students who were reading medicine.) Sometimes, the younger generation of scientists could make use of writings by the older generation. Maxwell, for example, found Brewster's book on *Optics* very helpful.

For mathematics, at least, a number of school textbooks were available. Babbage mentions one that attracted his attention:

> Among the books was a treatise on algebra, called 'Ward's Young Mathematician's Guide'. I was always partial to my arithmetical lessons, but this book attracted my particular attention. After I had been at this school for about twelvemonth, I proposed to one of my school-fellows, who was of a studious habit, that we should get up every morning at three o'clock, light a fire in the school-room, and work until five or half-past five.[24]

There was always the question of gaining access to relevant titles. Books were expensive – in the range of shillings or pounds each: a considerable amount when many, even of the middle-class, earned less than £200 a year. Efforts were made in the first half of the nineteenth century to produce cheaper educational books. Some of these attempts led to the appearance of cheap introductory texts on science. Henry Brougham, the politician, together with his friends formed the Society for the Diffusion of Useful Knowledge in the 1820s, which began to publish a Library of Useful Knowledge. Volumes appeared every fortnight at 6d. each. The series included a number of titles on scientific topics, ranging from brewing to the calculus. The authors were experts in their fields. John Daniel, for example, wrote on chemistry, and Peter Roget (of *Thesaurus* fame) on physics. Other publishers followed this example and produced series of cheap volumes covering both fiction and non-fiction. Thus the Chambers brothers in Edinburgh started publishing such miscellanies in the 1830s. Robert Chambers was particularly interested in science, and featured it in *Chambers's Journal*, which also started appearing in the 1830s. Spencer remembered reading this journal – and other periodicals, such as the *Mechanics' Magazine* – when he was a boy.

The advent of cheaper books and magazines certainly helped stimulate interest in science, but they still lay beyond the pocket of most schoolboys. Spencer had access to a range of books and periodicals because his father was honorary secretary of the well-supported Derby Philosophical Society, one of whose founders had been Erasmus Darwin, and which was subsequently supported by Lord Rayleigh's family. His father encouraged Spencer to use the resources of the Society's library. Similarly, Boole's father was a founder-member of the Lincoln Mechanics Institute, and this gave the young Boole access to its library. A number of fathers either borrowed books, or tried to form book collections of their own, which could then be read by their offspring. Wallace recalled reading books obtained by his father from reading clubs or subscription libraries. (Public libraries only became common in the latter half of the nineteenth century.) It was said of Airy that:

... he picked up a wonderful quantity of learning from his father's books. He read and remembered much poetry from such standard authors as Milton, Pope, Gay, Gray, Swift, etc. . . . But he also studied deeply an excellent Cyclopaedia called a Dictionary of Arts and Sciences in three volumes folio, and learned from it much about ship-building, navigation, fortification, and many other subjects.[25]

Like Airy, many scientists read widely when they were young. Shakespeare was a special favourite. Even Darwin – not a voracious reader as a child – was prepared to spend time reading Shakespeare's plays. Occasionally, such general reading was found to relate to scientific interests. Henry Armstrong always swore that he obtained most early assistance from a book entitled *On the study of words* by Richard Chevenix Trench, then Professor of Divinity at King's College, and later the geologist William Buckland's successor as Dean of Westminster.

Armstrong said it made him think for the first time about method and meaning. Similarly, Faraday was greatly influenced by *The improvement of the mind* by Isaac Watts (best remembered now as a hymn writer). William Grove, later a physicist, obtained his stimulus from fiction:

My first start was reading a child's story called the 'Ghost', where a philosophical elder brother cures his younger brother of superstition, by showing him experiments with phosphorous, electricity, etc. This set me on making an electrical machine with an apothecary's phial, etc. I was then about twelve years old.[26]

Backing from the family

The attitude of the family, and especially of parents, was often important. St. George Mivart, later a noted biologist, was grateful to his father for helping him despite the disapproval of others:

My father had no scientific knowledge: nevertheless, he encouraged me in all my tastes, giving me money freely for books and specimens, against the advice of friends.[27]

James Ewing, later Professor of Engineering at Cambridge, was the son of a Free Church minister in Scotland. He recalled that his parents allowed him to develop in his own way, though they had little interest in science:

In a family whose chief interests were clerical and literary I was a 'sport' who took his pleasure in machines and experiments. My scanty pocket money was spent on tools and chemicals. The domestic attic was put at my disposal. It became the scene of hair-raising and hair-singeing explosions. There, too, the domestic cat found herself an unwilling instrument of electrification and partner in various shocking experiences.[28]

Children whose family had an interest in science were even better placed. In this respect, Charles Darwin's home environment was excellent. His grandfathers – Erasmus Darwin and Josiah Wedgwood – had both been members of the Lunar Society in Birmingham, one of the powerhouses of science and technology in eighteenth-century England. Erasmus, though he made his living as a highly successful physician, was widely known as a larger-than-life polymath. Charles admired his grandfather, later in life writing a memoir about him. Wedgwood was an outstanding experimenter, who changed the nature of the ceramics industry. Charles' father was a medical man with an interest in research; his mother was a Wedgwood. So Darwin's interest in chemistry was not regarded as odd at home. A similar pattern can be found elsewhere. Lubbock's father, like Darwin's, was a Fellow of the Royal Society. The Lubbocks were near neighbours to the Darwins, and Lubbock, though a generation younger than Darwin, had similar interests. Darwin persuaded Lubbock's father that buying his son a microscope would be a worthwhile investment. Microscopes were expensive instruments: such a purchase reflects strong parental support.

Family attitude continued to be important further down the social scale. David Gill's father was a watchmaker, who encouraged his son to build a laboratory at home. Boole's father was a cobbler, but he was also interested in science and skilful with his hands. Father and son made scientific instruments together. The value of such encouragement was remarked on by Herbert Spencer:

> Most children are instinctively naturalists, and were they encouraged would readily pass from careless observations to careful and deliberate ones. My father was wise in such matters; and I was not simply allowed but encouraged to enter on natural history.[29]

In some cases, the mother played a significant role. Playfair's father was a senior medical officer in India, and so was absent from home for much of the time. Playfair's mother was well-known in St. Andrews, where they lived, and she was allowed to take her young son to science lectures in the university. Rayleigh's father had no links with science, but his mother's family had strong links with the Royal Engineers, an important body for the support of science in the nineteenth century. Roscoe's mother courageously encouraged him to carry out chemical experiments at home. Bentham's maternal grandfather was a Fellow of the Royal Society, and his mother aided her son's interest in botany. Lodge's mother was an enthusiastic photographer, though, in his case, it was an aunt who first encouraged his interest in science. As with Lodge, other relatives sometimes provided a stimulus towards science. Airy and Tait were both encouraged in their interests by uncles.

Even where families were less inclined to be supportive, they usually forbore from exercising a direct veto on scientific interests. Crookes' family thought he was a fool to be interested in chemistry, but they did not prevent him from experimenting. Though his mother supported his interest in science, Frankland was banned from doing chemical experiments by his step-father. (He got round this by using a hut lent to

A somewhat unkind cartoon of Herbert Spencer. Even sociologists rarely look this demonic. From *Vanity Fair* (26 April 1879). Newspaper Library, Colindale

him by friends.) Yet his step-father was clearly not unsympathetic, as the following account shows:

Whilst at this school, a very celebrated trial occurred in the adjacent assize courts of Lancaster Castle. It was an action brought by the Corporation of Liverpool against Mr. Muspratt for committing a nuisance by allowing muriatic acid gas to escape from his works in Liverpool. The action was tried before the late Baron Alderson, and as, at this time (fourteenth year of my age), I was already much interested in chemistry, my step-father allowed me to stay away from school in order to hear this trial. . . . Next morning the headmaster enquired why I had been absent the previous day. I told him the reason, he said 'that was no reason at all', and ordered me to commit to memory one hundred lines of Virgil before next morning. A letter from my step-father, however, procured for me the remission of this penalty.[30]

These examples suggest that the attitude of parents – and, to a lesser extent, other members of the family – was important in allowing children to develop an interest in science. Indeed, in a number of cases, the parents, more often the father, actively encouraged this interest. Such parents were unusual in the early years of the nineteenth century; though it is noticeable that some parental careers, most obviously medicine, seem to have helped develop a favourable attitude. But the range of family backgrounds of future scientists was large – from admiral to brewer, and from landowner to policeman. Most, however, were members of the rapidly expanding middle class.

The final conclusion must be that Charles Darwin's experiences, both at school and at home, as a boy interested in science typify quite well those of his contemporaries who were to become eminent scientists. Interest in science came early, though access to information about it was limited. School was often unhelpful. Family support, or at worst non-interference, was a significant factor in allowing the young boy to go ahead. Wealth helped – especially where, as with chemistry, apparatus and consumables were required.

⇥ *2* ⇤

Higher education

THE division between school and university was less clear-cut in the nineteenth century than it is today. Some future scientists were attending university while others of the same age were still at school. Scottish universities, in particular, registered students over a range of ages from around fourteen to the mid-twenties. Most of the scientists started at the younger end of this age scale. Lyon Playfair entered St. Andrews when he was 14 in the 1830s , and William Ramsay started at Glasgow at the same age in the 1860s. David Brewster went to Edinburgh even younger, at the age of 12, but the prize must go to Kelvin, who began his studies at Glasgow when he was only ten. Moreover, he won two class prizes in his first year, and continued to win prizes in every succeeding year. Scotland was not unique as regards early entry. J. J. Thomson, for example, entered Owens College [now the University of Manchester] when he was 14. But entrants to Oxford and Cambridge tended to be somewhat older. In some cases, studies at a Scottish university were the first step before a trip southwards to an English university.

Darwin at university

So Darwin was in no way exceptional when he started his studies at Edinburgh at the age of 16. Both his father and his grandfather had been to Edinburgh before him, and the intention was that he should become a medical man like them. Scottish universities offered a much wider curriculum than Oxford or Cambridge – in medicine especially, the practical training was better. Nevertheless, Darwin was not impressed by the teaching. The nadir from his viewpoint was the Professor of Anatomy, Alexander Monro, whose father and grandfather – both of the same name – had held the chair before him. 'I dislike him and his lectures so much that I cannot speak with decency about them,'[1] said Darwin. But his opinion of the rest of the teaching was not much better:

The instruction at Edinburgh was altogether by lectures, and these were intolerably dull, with the exception of those on chemistry by Hope; but to my mind there are no advantages and many disadvantages in lectures compared with reading.[2]

Thomas Hope was the Professor of Chemistry, and was one of the first to use large items of chemical equipment to demonstrate experiments to big audiences. The amount a Scottish professor earned depended in part on the number of students he could attract. Hope did very well: Darwin was one of some five hundred students to register for his class. Chemical demonstrations were all very well, but they were not the same as hands-on experiments. For this, geology at Edinburgh was better. Darwin thought that the lectures given by the professor, Robert Jameson, were very poor, but they were accompanied by three practicals a week in the museum – at that time one of the best geological museums in Europe. Besides providing ready access to specimens, Jameson also led field trips to examine the local geology – just the sort of outdoor practical activity that Darwin loved.

Darwin left Edinburgh without a degree. He was not unusual in this: taking a degree at that time was often a matter of personal inclination. Having decided against medicine, Darwin was now dispatched to Christ's College, Cambridge, with the idea of becoming a clergyman. At Cambridge, the basic subjects – classics and mathematics – were taught by tutors. The professors provided lectures on a range of subjects, but most of these were not included in the examination schedule. This discouraged students who wished to obtain high honours in their final degrees from attending, but allowed greater freedom in the nature of the teaching. Darwin, who was only concerned with taking a pass degree, initially took little interest in the lectures. One series that did attract him were given by John Henslow, the Professor of Botany (though his interests lay in natural history generally, rather than in botany alone). Darwin became so interested that dons in the university referred to him as, 'the man who walks with Henslow'. Darwin noted later that:

I attended, however, Henslow's lectures on Botany, and liked them much for their clearness, and the admirable illustrations . . . Henslow used to take his pupils, including several of the older members of the University, on field excursions, on foot or in coaches, to distant places, or in a barge down the river, and lectured on the rarer plants and animals which were observed. These excursions were delightful.[3]

The polymath William Whewell looking like the formidable debater he was.
From S. Douglas, *The life of William Whewell* (Kegan Paul, 1881).
The British Library, 4906.i.3

The 'older members of the University' that Darwin mentions included William Whewell, recently appointed Professor of Mineralogy, George Peacock, a leading mathematician, and Adam Sedgwick, Professor of Geology. Darwin subsequently attended Professor Sedgwick's geological lectures and pronounced them far more interesting than Jameson's in Edinburgh. His favourite reading, too, lay mainly outside the examination syllabus. He obtained particular encouragement from John Herschel's recently published *Preliminary discourse on the study of natural philosophy*, which dealt with the development of science, and the vast, seven-volume *Personal narrative* of Alexander von Humboldt, which described the German polymath's voyage to South America. Of the books that were officially accepted, Darwin enjoyed reading William Paley's theological works, not least, his *Natural Theology*. Despite his concentration on topics that did not figure in his examinations, Darwin managed to cram in enough knowledge of the required syllabus to obtain his B.A. in 1831. Indeed, he did quite well amongst those taking pass degrees: he came tenth out of nearly two hundred.

So, how does Darwin's student career compare with that of other scientists. What universities did they go to (if any)? What subjects did they study? What was the teaching like? What books did they read? The first of these questions can be answered in general terms as follows. In the first few decades of the nineteenth century, choice of universities was limited. It was usually a question of going either to one of the four universities in Scotland, or to one of the two in England. (Trinity College Dublin catered mainly for Irish students.) New universities and colleges in England started to appear in the 1830s, but their student intakes only grew slowly. In terms of university attendance, we can therefore divide our scientists into three main groups – those who went to Oxford or Cambridge, those who went to other universities, and those who did not attend university at all. People in this last group acquired an interest in science, and a knowledge of it, in a remarkable variety of ways. One sub-set of the group includes officers in the Army or Navy. The Army set up the Royal Military Academy – universally known as 'the Shop' – at Woolwich in the mid-eighteenth century. Its teaching covered a fair amount of mathematics, as well as surveying. A number of the officers trained there, especially those who went into the Engineers or the Artillery, subsequently made an impact on nineteenth-century science. Naval officers did not acquire an equivalent establishment until later in the nineteenth century. However, on-the-job training in navigation and cartography provided several with an incentive to take an interest in science. Robert Fitzroy, with whom Darwin sailed in the 1830s, comes immediately to mind, and others will appear in due course. But this chapter will concentrate on those scientists who, like Darwin, went on to some form of higher education.

Oxford and Cambridge

In the first half of the nineteenth century, both Oxford and Cambridge aimed at educating people who would enter the traditional professions (medicine, the law and the Church), along with a considerable number of well-off young men who were there mostly for the experience. Training for the Church was given especial emphasis, both because the Fellows of the colleges, where most of the teaching went on, were normally ordained clergyman of the Church of England, and because the colleges controlled the appointment of clergy to many benefices up and down the country. Where the two universities differed was in the main thrust of their teaching. Oxford emphasised the classics; Cambridge emphasised mathematics.

This difference mainly affected the brightest students, who wished to demonstrate their intellectual prowess by taking a high honours degree. At Cambridge, they had to show outstanding ability at mathematics – something that attracted young men with an interest in physical science. The course at Cambridge culminated in an intensive written examination for what was called the Mathematical Tripos. The students' results were placed in the order of their marks. The one who came head of the list was referred to as the Senior Wrangler; he was followed by the Second Wrangler, and so on. The top students were allowed to compete for the Smith's Prize examination, which followed the Tripos examinations. Doing well in the Tripos and the Smith's Prize (there were actually two of them) meant the almost automatic offer of a college fellowship. The roll call of students falling into this category includes many of the great names of nineteenth-century science. In 1813, John Herschel was Senior Wrangler and First Smith's Prizeman, with George Peacock – the same man who attended botany classes with Darwin – as Second Wrangler and Second Smith's Prizeman. Ten years later, Airy emulated Herschel's success, as did Arthur Cayley in the 1840s, Tait in the 1850s, and Rayleigh in the 1860s. In the early 1850s, Maxwell was Second Wrangler and First Smith's Prizeman, as was Kelvin ten years earlier. Kelvin was actually quite sure he would be Senior Wrangler. The story runs that when the examination results were put up, Kelvin sent his servant to look at the list. When the servant returned, Kelvin asked him, 'Who came second?'. The servant replied, 'You did, sir'.

Until the latter half of the century, undergraduates at Oxford and graduates at Cambridge were required to give assent to the Thirty-Nine articles of the Church of England. At Cambridge, this ruled out Jews, such as James Sylvester, who was Second Wrangler in his year, and many, though not all, Non-conformists from receiving their degrees. There were exceptions, like Isaac Todhunter (Senior Wrangler in 1848). He came from a Congregationalist background, but was prepared to accept the Thirty-Nine articles. Augustus De Morgan (Fourth Wrangler) was a member of the Church of England. Typically, he refused to assent to the articles on the grounds that he did not believe in religious tests. When religious tests at Oxford and Cambridge were finally abolished after the mid-century, some of those previously disbarred – including Sylvester – finally took their degrees. With a lack of logic customary in old universities, honorary degrees were not subject to the same rule. So when the British Association met in Oxford in 1832, honorary degrees were awarded to four eminent scientists – Dalton, Brewster, Faraday and Brown (a leading botanist) – to celebrate the event. None of them was a member of the Church of England.

During the nineteenth century, the topics covered in the Mathematical Tripos gradually changed – mainly due to the efforts of the leading Wranglers. When John Herschel went up to Cambridge in 1809, mathematical in England was in the doldrums. Continuing devotion to old-fashioned approaches meant that the subject had lagged further and further behind the Continent during the eighteenth century. Herschel, along with his friends Babbage and Peacock, set up the Analytical Society at Cambridge. As its name indicated, their intention was to try and import the new continental mathematics into England. Though Herschel and Babbage left Cambridge, Peacock stayed on, and became an examiner for the Tripos. With help from others, such as Whewell, Peacock introduced continental mathematics into the examinations in the 1820s. In subsequent years, Airy and Whewell further broadened the topics covered in the Tripos, mostly by including areas of physics. These changes certainly made the examination more interesting for those with an inclination towards physical science, but J. J. Thomson, looking back over the nineteenth century, was dubious about the overall approach followed in the Tripos: 'Whether or not the changes that have been made in the Mathematical Tripos have been instrumental in promoting the progress of mathematical science in this country is a question which it is very difficult to answer'.[4]

Success in the Tripos depended, in part, on going to the right coach. Earlier in the nineteenth century, the great name amongst the coaches

The astronomer John Herschel in his later years, by which time his bushy hair style had become a characteristic feature.
The British Library, 10604.g.10

was William Hopkins. He produced 17 Senior Wranglers down the years, and his pupils included Stokes, Cayley, Kelvin, Tait and Maxwell. These coaches were private tutors who had to be paid. Airy noted that, when he started taking pupils, he charged them £14 per term, and a popular coach could get appreciably more than that. But Tait's explanation of what coaches did goes some way to explaining J. J. Thomson's doubts about the Mathematical Tripos. They were always:

eagerly scanning examination papers of former years, and mysteriously finding out the peculiarities of the Moderators and Examiners under whose hands their pupils are doomed to pass, spend their lives in discovering which pages of a textbook a man ought to read and which will not be likely to 'pay'. The value of any portion as an intellectual exercise is never thought of; the all-important question is – Is it likely to be set?[5]

Later in the century, Edward Routh became the leading coach. Routh had been one of Hopkins' pupils, and was Senior Wrangler in the year when Maxwell was Second Wrangler. He beat Hopkins' record by producing 27 Senior Wranglers over a period of 33 years. Unlike most coaches, Routh taught via lectures, rather than tutorials, but his approach was basically similar. As J. J. Thomson recalled it:

In his lectures he took us through the best textbook on the subject, the parts which the author had treated satisfactorily he just told us to read; when the book was obscure he made it plain; when the proof of a theorem was longer than need be he gave us the shortest one; when the author had put in something that was not important he told us not to read it; when he had omitted something important he supplied the omissions.[6]

Success in the Tripos depended on speed, as well as on mental ability. Rayleigh reported that, because his name came late in the alphabet, he was able to complete an additional question while the papers of earlier candidates were being collected. Good health was equally important. The examinations were held in the winter, and there was no heating in the examination hall. This certainly affected some people:

A man with any tendency to imperfect circulation becomes chilled, especially in his hands, and with chilled hands, he is disabled to a considerable extent from writing. The first year I was at Cambridge, one of our best Trinity men, afterwards a Fellow, lost fifteen or eighteen places among the Wranglers, as he believed, and as previous and subsequent successes entitled him to believe, solely from being frozen up.[7]

In the early years of the century, teaching frequently involved manuscript material prepared by the individual tutor. Airy recalled copying out some of Whewell's original manuscripts on such matters as tides and the shape of the Earth. These were supplemented by a limited range of textbooks in English along with Newton's *Principia* in Latin. As time passed, other texts became important. The Analytical Society's agenda emphasised particularly the importance of understanding developments in France. In consequence, the keenest mathematics students began to read some of the key French writings. Airy remembered struggling with the French vocabulary as he tried to read Poisson's *Mechanics*. Later, when Maxwell listed the books he read, his top titles included books by the French mathematicians Fourier, Monge and Cauchy, as well as Poisson. Such books were not cheap. Maxwell had to pay 25 shillings for his copy of Fourier's *Theory of Heat* – a week's wage for many people. He considered it money well spent, calling the book a great mathematical poem (an opinion with which Kelvin agreed).

By the mid-century, the older generation of British scientists and mathematicians had provided texts of their own. In the 1860s, Rayleigh noted that he had read books by Whewell, Herschel, Babbage and Boole, but half of his mathematics reading was of textbooks by the remarkably productive Isaac Todhunter. Todhunter was a Fellow of St. John's College, and one of the few college tutors who could compete effectively with the leading private tutors. In fact, his books, which went through many editions, did sufficiently well financially for him to relinquish his Fellowship and to get married. But as Rayleigh also noted, the Cambridge mathematical tradition did not help in developing the training of experimentalists. When he wanted to learn experimental techniques, the only hands-on course he could find in Cambridge was one on qualitative chemical analysis. Doubting the value of experimental work for undergraduates was common enough in Cambridge. It was expressed most memorably by Todhunter:

It may be said that a boy takes more interest in the matter by seeing for himself, or by performing for himself . . . this we admit, while we continue to doubt the educational value of the transaction. The boy would also probably take much more interest in football than in Latin grammar . . . It may be said that the fact makes a stronger impression on the boy through the medium of his sight, that he believes it the more confidently. I say that this ought not to be the case. If he does not believe the statements of his tutor – probably a clergyman of mature knowledge, recognised ability and blameless character – his suspicion is irrational, and manifests a want of the power of appreciating evidence, a want fatal to his success in that branch of science which he is supposed to be cultivating.[8]

Though both Oxford and Cambridge introduced science degrees in the 1850s, ambitious Cambridge-produced scientists of the nineteenth century still typically took a high honours degree in the Mathematical Tripos. This did not require an undergraduate to do experimental work (laboratory teaching only began to flourish at Cambridge in the 1870s). Compare this with Darwin's pass degree, and his extensive practical experience. An eminent Cambridge physicist of the twentieth century divided the sciences into two groups – physics and stamp collecting. The Wranglers clearly belong to the first group; Darwin to the second. Consequently, he cannot be considered a typical Cambridge scientist of the period. His university career is better compared with that of his friend, Charles Lyell, who was one of the few eminent scientists to go to Oxford.

Lyell took a respectable honours degree in classics, but he had become interested in geology by reading Bakewell's *Geology*, found in his father's library. (Robert Bakewell was a significant figure in British geology round the turn of the century. He suffered from being constantly confused with a fellow-Midlander of the same name who pioneered new methods of farming. 'Tell me', gushed a Countess to the geologist, 'are you related to the man who invented sheep?'.) His reading led Lyell to the lectures given by the leading geologist in Oxford, William Buckland. In the early decades of the century, Oxford was even better known for geology than Cambridge, and the able, eccentric Buckland was the root cause. He promoted fieldwork enthusiastically, and was a distinctive figure as, clad in a top hat, frock coat and umbrella, he led his students round the countryside. According to Lyell, his lecturing methods were unusual:

[He] would keep his audience in roars of laughter, as he imitated what he thought to be the movements of the *Iguanodon* or *Megatherium*, or seizing the ends of his clerical coat-tails would leap about to show how a pterodactyl flew.[9]

Buckland was a great believer in the importance of taste. Students and colleagues had to be prepared for unusual meals when they visited him. Buckland, himself, thought that the most unpleasant things he had ever eaten were moles and bluebottles. His son, Frank, who became a well-known (if also somewhat eccentric) nineteenth-century naturalist, recalled visiting a cathedral on the Continent with his father. Dark spots on the pavement, apparently always refreshed, were shown to them, and claimed to be the blood of a martyr. The elder Buckland

The geologist William Buckland geared up to inspect a glacier (though his headgear reminds us that he was still a Victorian gentleman). From A. Geikie, *The life of Sir Roderick I. Murchison* (John Murray, 1875).
The British Library, 10825.dd.14

dropped on his hands and knees and licked one of the spots. After mature consideration, he diagnosed bat's urine. Darwin at Cambridge, where they took things more seriously, thought Buckland was a vulgar man who craved notoriety.

Ireland, Scotland and elsewhere

Lord Rosse was at Oxford at much the same time as Lyell, but he concentrated more on mathematics. He had come via Trinity College Dublin, an Elizabethan foundation set up on the lines of an Oxford or Cambridge college, and catering primarily for the Protestant community in Ireland. At the beginning of the nineteenth century, Trinity College Dublin had no great reputation in science, though this changed as the century progressed. Robert Ball, subsequently a leading Irish astronomer, went there at the end of the 1850s, having just become interested in astronomy as a result of his reading:

Shortly after leaving school I was given a copy of an introduction to astronomy by Mitchell, which is known by the name of 'The Orbs of Heaven'. I well remember sitting up to the small hours of the morning devouring this book. It delighted me as few books have ever done before or since.[10]

At Trinity College Dublin, he found that, along with mathematics, there was a compulsory course in astronomy, which provided him with the basics that he wanted. The textbook for the course was Brinkley's *Astronomy*. This was a home-grown product, in the sense that Brinkley had been Professor of Astronomy in Dublin earlier in the century. (He had resigned the chair on his appointment as Bishop of Cloyne. It was said in Ireland that he should have thanked his stars for his promotion.) In later years, Ball had to tackle much more difficult texts: not only Newton's *Principia*, but also Laplace on *Celestial Mechanics*.

Another Irishman, George Stokes, went straight to a college in Bristol. This had a principal who was also Irish: a graduate of Cambridge whose course provided an excellent introduction to what was required of undergraduates there. Stokes, like others from Ireland who went to Cambridge, valued not only the specialist training provided by the Mathematical Tripos, but also the prestige and career opportunities that it offered. Tait used Edinburgh University in much the same way as Stokes used his college in Bristol. He spent one year there, concentrating on mathematics and natural philosophy; then he went on to Cambridge. His friend, Maxwell, stayed three years at

The astronomer Robert Ball. Though a competent scientist, he was better known to his contemporaries as a populariser of astronomy. From *Vanity Fair* (13 April 1905).
Newspaper Library, Colindale

Edinburgh, and Kelvin, a decade earlier, for six years at Glasgow, before they moved on to Cambridge. (Neither of them took Scottish degrees.)

Maxwell's record of a typical day at Edinburgh runs as follows:

I look over notes and such till 9:35, then I go to Coll. . . . at 10 comes Kelland. He is telling us about arithmetic, and how the common rules are best. At 11 there is Forbes, who has now finished introduction and properties of bodies, and is beginning Mechanics in earnest. Then at 12, if it is fine, I perambulate the Meadows; if not, I go to the Library and do references. At 1 I go to Logic. . . . Then I extend notes, and read text-books, which are Kelland's *Algebra* and Potter's *Mechanics*. The latter is very trigonometrical, but not deep.[11]

The distinctive feature here is the emphasis on philosophy and logic in the Scottish curriculum. Despite their transition to Cambridge, both Maxwell and Kelvin regarded this greater breadth as being valuable. Kelland, mentioned by Maxwell, was the Professor of Mathematics at Edinburgh, while James Forbes was the Professor of Natural Philosophy. In those days, science tended to be divided into natural philosophy and natural history – roughly speaking, a division between the physical sciences and the biological sciences. In the early nineteenth century, chemistry was beginning to blossom as a subject in its own right, so 'natural philosophy' increasingly meant physics. Forbes had, himself, attended natural philosophy courses at Edinburgh in the 1820s, and remembered the lectures with more pleasure than Darwin did:

Dr. Jamieson, plain, practical, not to say prosaic, but accurate, painstaking, and diligent as an observer, so rarely ventured on figures of speech that the one or two metaphors in which during the whole session he indulged were well known and waited for, and when produced were welcomed with annual rounds of applause.[12]

Forbes was not exposed to hands-on work as a student, nor did he provide practical classes for his own students. Maxwell was made an exception: Forbes had constructed a laboratory for himself and gave Maxwell the run of it. As Darwin found, practical classes were rarely available in Scotland. Playfair chose to go to the Andersonian College in Glasgow in the 1830s, because there, Thomas Graham (who had studied in Edinburgh under Darwin's favourite lecturer, Hope) provided both lectures and a laboratory. Later in the century, Playfair commented that to Davy and Faraday, 'we owe largely the popularity of chemistry in this country; but none of these chemists, except Graham, thought of opening their laboratories for the training of students in the methods of research.'[13] Practical training in laboratory work gradually became commonplace in the nineteenth century, mainly due to the efforts of Playfair's contemporaries and their successors.

A significant change in the nineteenth century was that the openings for higher education expanded considerably. The first new institution to appear was University College London in the 1820s. Much of the impetus behind its creation came from Scotland, and the syllabus it offered was broad, even by Scottish standards. Equally, it had no religious tests. Amongst its first professors were De Morgan and Sylvester. This 'godless institution in Gower Street', as Thomas Arnold called it, was soon matched by an Anglican foundation in London – King's College – which provided an equally wide range of subjects, but now plus religious instruction. A few years later, in the mid-1830s, the University of Durham was established, mainly on the basis of money provided by Durham Cathedral. Needless to say, it was an Anglican foundation. Then, at the mid-century mark, Owens College was founded in Manchester. If King's College London and Durham shared conservative and Anglican sympathies, University College London and Owens College shared more liberal and Nonconformist attitudes.

The problem for several decades was that all of these institutions had financial problems, not least because they found it difficult to recruit students. In some cases, the institution simply acted as a stepping-stone for entry to Oxford or Cambridge. Francis Galton, for example, studied medicine and chemistry at King's College London before going on to take mathematics at Cambridge (partly on the advice of his cousin, Charles Darwin). He was not alone in looking to London for medical training. For many years, the important part of University College London was its hospital, and the college was dominated by medical students. This was not always regarded with pleasure by college staff. As one remarked concerning the teaching of reproduction in the biology classes, this would, 'inevitably corrupt the youthful mind. It would make all students medical students'.[14] By the standards of the time, University College provided a good medical education. Joseph Lister, who entered in 1843, noted that he observed in University College Hospital the first operation using ether in the British Isles.

Traditional reasons for studying

In looking for future careers, most undergraduates considered the traditional professions. It raised few eyebrows in the 1830s, when

Charles Darwin went to Cambridge with the thought of entering the Church. His father saw it as a reasonable option if he would not do medicine: after all, various of their relatives were clergymen. From Darwin's viewpoint, some of the people he came to admire at Cambridge, including Henslow, were members of the clergy. Leonard Jenyns, Henslow's brother-in-law, was the incumbent of a parish near Cambridge. He was an avid naturalist, and Darwin enjoyed visits to his parish, where they went collecting together. So, as Darwin came to the end of his time at Cambridge, a future as a parson-naturalist seemed to beckon.

In the first half of the nineteenth century, Fellows of colleges at Oxford or Cambridge were typically ordained, or on the way to ordination. They were expected to be celibate, though professors and heads of colleges were allowed to marry. Whewell's career at Cambridge illustrates the sequence nicely. He was elected a Fellow of Trinity in 1817, ordained in 1825, became Professor of Mineralogy in 1828, and married in 1841. As this illustrates, it was possible to be a Fellow for a number of years without being ordained, but the regulations for remaining permanently in a college were more stringent. These requirements deterred some. Equally important was a new seriousness that affected religious thinking in the early decades of the nineteenth century. In the previous century, it was acceptable practice to send a younger son into the church, where he could, if he wished, spend a good deal of his time hunting, fishing and shooting. This became less and less acceptable as the nineteenth century progressed. Darwin and his contemporaries came in at the tail end of the old tradition.

Considering the church as a option was not limited to undergraduates at Oxford and Cambridge. Brewster at the beginning of the century long believed that he was destined to become a minister in the Church of Scotland. Indeed, he became a licensed preacher. As one of the congregation later recollected:

Brewster preached his first sermon in . . . the West Kirk of Edinburgh, one of the largest in Scotland, accommodating 2500 hearers. This spacious church, with its double tier of galleries, was on this occasion unusually crowded, for the reputation he had acquired drew together, not only numbers of his fellow-students, but of literary and scientific men, anxious to hear how he would begin his professional career.[15]

Some years later, his fellow-Scotsman, James Forbes, inclined towards the Church of England. He explained in a letter to an uncle:

Perhaps it may be news to you that my earliest wish towards a profession was for the Church of England; when I was eight years old, I began to compose sermons, and long before my thoughts turned to science I was warm on the scheme of the Church. In November 1822, I seriously proposed becoming a clergyman, but was dissuaded from it by my family. . . . Law was the choice of my father, and, on the principle of extinction of others merely, my own. My distaste increased instead of diminishing. Indeed I could not have tolerated the idea of such a struggle of mind, but in the hopes of after a few years of drudgery reaching a sheriffship, which by adding £300 or £400 to my income might enable me to pursue my darling studies at leisure.[16]

Forbes was not alone in finding legal studies distasteful. John Herschel's father believed that his son should become a clergyman – for much the same reasons that Darwin's father recommended that career to his son. John did not agree and, against his father's advice, decided to go into the law instead. He soon regretted his choice. Whewell, at about that time, wrote to a friend considering a legal career: 'Your profession has a greater tendency than others to efface the simplicity and energy of the mind'.[17] John Herschel soon came to have much the same feelings. Nevertheless, some eminent scientists did enter the legal profession successfully. The two mathematicians and close friends, Cayley and Sylvester, both entered the Inns of Court in London, and would walk round Lincoln's Inn together discussing their current mathematical interests. In fact, a number of lawyers who were subsequently eminent in their profession took the Mathematical Tripos at Cambridge, which may explain some of the attraction of the legal profession to scientists who took the Tripos. It was not, however, an essential background. William Grove graduated from Oxford with a pass degree, and entered Lincoln's Inn. He seems to have been satisfied with the law as a career, but ill-health prevented him from practising it much, while giving him ample time to concentrate on science. His combined legal and scientific background proved very useful when the Royal Society decided to revise its Charter in the mid-nineteenth century.

Of all the traditional professions, medicine was the one that had the most immediate contact with science, especially via anatomy and physiology. Traditionally, surgeons were regarded as a lower level of practitioner. They were trained on the job, whereas physicians were more likely to have an Oxford or Cambridge background. In the early decades of the nineteenth century, this distinction was beginning to break down. One indication of the change was the appearance of a new hybrid medical man – the surgeon-apothecary. As the name suggests, such practitioners could handle both the medical and the surgical aspects of a doctor's work. They underwent a lengthy training, often requiring a more scientific attitude to medicine than had been common in the past, and were correspondingly more open to change. So Davy began his career as an apprentice to a surgeon-apothecary in Penzance who encouraged his interest in chemistry.

To obtain a medical training, some, like Darwin, went to study in Scotland; but the main centre of attraction was the London teaching hospitals. Richard Owen went to Edinburgh for half a year, but then moved to London. Thomas Huxley, the man who was to be his main scientific rival in later years, went to directly to London – to Charing Cross Hospital – though he was not impressed by the teaching that was provided. The only exception was physiology: the 'mechanical engineering of living machines', as Huxley called it.

the only teaching from which I obtained the proper effect of education was that which I received from Mr. Wharton Jones, who was lecturer on physiology at the Charing Cross School of Medicine. The extent and precision of his knowledge impressed me greatly, and the severe exactness of his method of lecturing was quite to my taste. I do not know that I have ever felt so much respect for anybody as a teacher before or since.[18]

New openings

By the mid-century, new opportunities in science were slowly beginning to open up. Chemistry, in particular, was seen to have some commercial value. Obtaining a chemical training became a little easier after 1845, when the Royal College of Chemistry was set up in central London. An eminent German chemist, August Hofmann, was imported to run it, and, although its finances remained shaky, it produced a string of important scientists. Not all continued into chemistry. For example, Warren de la Rue, who attended the college right at the start, made his name in astronomical photography. But many of the students did make their names in chemistry. William Crookes joined in 1848, just as the Chartist riots were filling the news. William Perkin was studying there in the 1850s, when he made his breakthrough in the manufacture of artificial dyes. Two decades later, his son William Perkin Junior, also went to the College. Meanwhile, in the mid-1860s, Frankland took over from Hofmann, who had been appointed to a chair in Germany. Henry Armstrong entered the College just as Hofmann was leaving. He had had no previous experience in chemistry and revelled in the opportunities now offered to him, becoming so interested in reading the books available that he was made College librarian. Frankland's teaching was very much to his taste: 'His lectures were clear, straightforward and logical and he took particular pains to illustrate them by well-thought-out, practical demonstrations'.[19]

The Royal College of Chemistry offered certificates and diplomas, rather than degrees, but many of the students only remained for a brief time in order to pick up specific knowledge or skills. Students were of all ages and backgrounds. George Liveing came down from Cambridge for a month in order to learn how to run a practical course in chemical analysis. (This was the sole practical course that Rayleigh found when he was at Cambridge.) Many others came from industries requiring some chemical knowledge, such as brewing, to try and explore new techniques or ideas.

The College only operated as an independent body for a limited period of time. In 1853, it was united with the recently created Government School of Mines. In effect, the former was seen as a place for teaching practical chemistry, while the latter was a place for teaching practical geology. The combined institution came under the control of a newly created governmental body – the Department of Science and Art, or, as Huxley called it, the 'Science and Tarts' Department. By 'Tarts', he was probably not casting doubts on the Department's morality – that meaning was only just coming in. He meant that it dealt with things that had little theoretical basis, like the baking of tarts. For the word 'Art' in the title referred to the practical arts – making the kind of things from furniture to technical gadgets which came to be displayed (as they still are) in the Victoria & Albert Museum. Playfair headed the scientific side of the new Department, which, as the century drew on, came to involve many of the country's leading scientists in its work. One particularly important task was the training of science teachers. Oliver Lodge combined attending evening lectures at King's College London with taking the course for teachers laid on by the Science & Art Department at South Kensington. The latter had the advantage that participants received some funding. Lodge explained:

The arrangements were that the successful applicant had to go up to London, keep regular hours in the laboratory, signing a book every day when he arrived before ten o'clock in the morning, and receiving for maintenance thirty shillings a week, handed over weekly by the clerk in charge of the book. At ten o'clock the clerk drew a red line, and if you signed below that line you were fined a certain amount.[20]

This question of money was one that greatly affected how, or whether, a scientist could pursue his studies. At one end of the scale, Rayleigh entered Cambridge as a fellow-commoner, which meant that he had extra privileges, such as dining with the dons rather than with the other students. At the other end, Boole could not afford to go to Cambridge at all, and George Green, a Nottingham miller who became one of the great mathematicians of the nineteenth century, could only afford the time and the money at the age of 38. Some idea of the amount involved can be obtained from Kelvin's experience. He estimated that the total annual cost of maintenance at Cambridge in the 1840s was £230.7s.8d. – perhaps £13,000 a year in modern terms. As Kelvin noted, college and university prizes helped in paying for books, which were an expensive item. Playfair benefited from this prize system in an unusual way:

In one of my annual trips to Arran, in 1836, I carried with me the first prize of the chemistry class. The book was a handsomely bound copy of Lyell's *Geology*, which I read on my way down the Clyde. A charming lady sat next to me in the steamer. We entered into conversation, and she asked me the name of the book which interested me so much. I explained to her that Lyell, the author of the book, had established geology on a new basis . . . The lady seemed to be amused, and said that she was glad Lyell had such an enthusiastic admirer, because she was his wife, and that gentleman on the other side of the boat was Lyell himself. My hero-worship had its reward, for she beckoned to the great geologist and introduced me to him. . . . This was the basis of a friendship which lasted till the death of Sir Charles and Lady Lyell.[21]

Darwin and others

But what of the original question – how typical was Darwin's university career? There was nothing unusual in attending more than one university, as he did. A number of eminent scientists, especially those from Scotland, did the same. At Cambridge, Darwin was unusual in taking a pass degree and in concentrating on practical natural history. Most scientists at Cambridge who were subsequently eminent took an honours degree in mathematics with no associated practical work. Even the more liberal syllabuses of the Scottish and the new English universities did not include much practical work up to the mid-century. In the latter half of the century, this changed, and the importance of practical training came to be accepted. From that viewpoint, Darwin's university career was more nearly the norm for his successors, rather than for his contemporaries.

Poor lecturing was commonplace. Outside Oxford and Cambridge, academic salaries depended, in part, on attracting large numbers of students. Introductory courses, which were most likely to attract students, were therefore often presented reasonably well. At Oxford and Cambridge, little by way of science appeared in the examination syllabus. So professors of scientific subjects there had considerable freedom in the way they presented their material. This freedom might lead them to ignore their duties, as Babbage did. He was professor of mathematics at Cambridge for eleven years, but never gave a lecture. Equally, it might lead them to present their subject in an innovative way – as happened with Henslow. In general, teaching improved as the century progressed. Even so, reluctant lecturers – such as James Dewar, who was appointed professor at Cambridge in 1875 – could still be found. The most effective teaching was in mathematics, and especially in the tutorial classes at Cambridge. The problem with them was their concentration on examination questions. Most scientists agreed with Darwin that reading round their subject was important. In the first half of the century, the ability to do this was restricted by the rather limited amount of relevant reading matter available in English. So Darwin's contemporaries often read books imported from the Continent. Later, these same contemporaries wrote books that helped the next generation of scientists.

⤙ *3* ⤚

Training and research

Not long after he had finished at Cambridge, Darwin received a letter from his former professor there, Henslow, backed by another from Peacock. They urged him to accept the post of naturalist on a round-the-world trip that was currently being planned. Initially, Darwin's father was opposed to the idea. He thought it was yet another example of his son putting off the choice of a proper career. But he was eventually persuaded to back the plan. There was a further hurdle: Darwin had to survive the scrutiny of Robert Fitzroy, the captain of the ship involved, who was looking for an acceptable companion to take on board. Fitzroy had exacting social standards: he was, after all, a descendant of Charles II and related to several of the leading aristocratic families in the country. Darwin's background and sponsors proved satisfactory, and he was duly appointed the naturalist aboard Fitzroy's ship, the *Beagle*. Though Darwin had his training at university, his books, and his correspondence with friends to help, his initiation into a research career depended essentially on self-teaching during his five-year voyage.

Travel and training

A trip abroad was, indeed, the way of gaining research experience for a number of eminent Victorian naturalists. With many areas of the world still unexplored, they could make names for themselves by identifying and bringing back new flora and fauna. The great name in natural history at the beginning of the nineteenth century was Joseph Banks, President of the Royal Society for 42 years up to his death in 1820. Banks had established his reputation by accompanying Captain James Cook on his voyage to the South Seas in the *Endeavour*, and was seen as providing a model for later naturalists. After his return, he helped establish the Botanic Gardens at Kew, where William Hooker became Director in 1841 (to be followed some years later by his son, Joseph). Equally important as a model was the German, Alexander von

The Director of Kew Gardens, Joseph Hooker, in contemplative mood.
The British Library, PP1931pch

Humboldt, who spent five years around 1800 studying the natural history and geology of South America. His writings influenced both Darwin and Wallace.

Joseph Hooker's family was not as wealthy as Darwin's, but his father – then Professor of Botany at Glasgow University – made up for that by his range of useful contacts. One of them was the naval explorer, Captain James Ross, who was commissioned to explore the Antarctic at the end of the 1830s. He offered to take the younger Hooker with him on his ship, the *Erebus*. In this case, Hooker could help pay his way, since, unlike Darwin, he was medically qualified. So he made the voyage both as naturalist and as assistant surgeon. For Huxley, who had neither wealth nor good contacts, embarking on a voyage was less straightforward. What he did have was good medical qualifications. As he later explained:

It was in the early spring of 1846,that, having finished my obligatory medical studies . . . I was talking to a fellow-student, and wondering what I should do to meet the imperative necessity for earning my own bread, when my friend suggested that I should write to the Director-General for the Medical Service of the Navy, for an appointment. . . . Having passed this, I was in Her Majesty's Service, and entered in the books of Nelson's old ship, the *Victory*, for duty at Haslar Hospital. My official chief at Haslar was a very remarkable person, the late Sir John Richardson, an excellent naturalist and far-famed as an indomitable Arctic traveller. . . . After a long interval, he stopped me, and describing the service on which the *Rattlesnake* was likely to be employed, said that Captain Owen Stanley, who was to command the ship, had asked him to recommend an assistant surgeon who knew something of science; would I like that? Of course I jumped at the offer.[1]

For Wallace, who left school at the age of thirteen, progress was more difficult still. His brother trained him as a surveyor, and, Wallace, as he carried out successive surveys, began to develop an interest in geology and botany. When surveying work became harder to find, Wallace spent a period as a school-teacher in Leicester, where he met a keen local naturalist, Henry Bates. The plans for new railways in the 1840s led to a renewed demand for surveyors, and Wallace returned to surveying. The salary of two guineas a day that he received for this was princely compared with the £30–40 a year plus accommodation that he had been earning as a teacher. From it he saved enough to contemplate a collecting trip abroad, during which he might develop his credentials as a naturalist. He planned to make two collections: one to study and describe on his return, and one to sell in order to finance his trip and his studies. Bates agreed to come with him, and, by 1848, they had agreed where to go. The difference in their approach from Darwin's is very evident:

What decided our going to Para and the Amazon rather than to any other part of the tropics was the publication in 1847 of *A voyage up the Amazon* by Mr. W. H. Edwards. This little book was so clearly and brightly written, described so well the beauty and the grandeur of tropical vegetation, and gave such a pleasing account of the people, while showing that expenses of living and of travelling were both very moderate, that Bates and myself at once agreed that this was the very place for us to go to if there was any chance to pay our expenses by the sale of our duplicate collections. We immediately communicated with Mr. Edward Doubleday, who had charge of the butterflies at the British Museum, for his advice upon the matter. He assured us that the whole of Northern Brazil was very little known, and that there was no doubt we could easily pay our expenses. Thus encouraged, we determined to go to Para, and began to make all the necessary arrangements.[2]

Naturalists were not the only people who foresaw a scientific benefit from travelling: so did geologists. Lyell went on an extended trip to the Continent after he had finished at Oxford. This was, in part, a Grand Tour of the type that young men from well-off families had traditionally made. But Lyell also used it to extend his knowledge of geology. His contemporary, James Forbes, went on a similar tour, and was much impressed by Vesuvius, writing a paper about it when he returned. Henry de la Beche intended to become an Army officer, like his father. But the end of the Napoleonic Wars in 1815 put paid to this idea, while, at the same time, making it easy for tours of the Continent to start again. So de la Beche spent some time doing geological fieldwork on the Continent instead of entering the Army. Roderick Murchison left the Army when the war ended, but did not immediately turn to geology. Instead, he preferred hunting and shooting. One of the people he shot with was Humphry Davy, who persuaded him that he had the qualities needed to become a geologist:

He saw that as a sportsman I had a quick and clear eye for a country, and that with most mountain forms and features I was already acquainted, and so he stimulated me to sell my horses, settle for the winter in London, and attend the chemical and mining lectures of the Royal Institution.[3]

Murchison learnt his science from the Royal Institution at the top end. Michael Faraday came in at the bottom. Faraday, with no money and no contacts, faced great difficulties in obtaining scientific experience. He

The Scottish scientist James Forbes with a companion, studying a glacier in the Alps.
The British Library, 1429.K.7

Michael Faraday, chemist and physicist, around the middle of the nineteenth century. The object in his hand is a scientific instrument, not a cigar.
The British Library, 10804.f.6

had only a limited amount of schooling, and was fortunate to be apprenticed at the age of fourteen to a London bookseller. Here he began to read in earnest. He was instructed in the art of bookbinding, and took the opportunities this gave to read books that attracted him. Binding the *Encyclopaedia Britannica* led him to an article on electricity, and he began to experiment with homemade electrical apparatus. Binding chemistry books led to chemical experiments in a small area he was allowed to use at the back of the shop. Faraday's interest in science came to be known, and a friend of the bookseller gave him a ticket to attend Davy's lectures at the Royal Institution. When Davy injured an eye in an explosion, Faraday, who had beautiful handwriting, was recruited to take notes for him. Shortly afterwards, Davy's assistant was sacked for brawling and Faraday was offered the job. (His salary was only a guinea a week plus accommodation, but he had continuous access to the laboratories.) Even Faraday benefited from travel abroad. Soon after his appointment, he accompanied Davy on a Grand Tour of Europe. Admittedly, he went in a menial capacity as Davy's valet, but it still allowed him to meet some of the leading scientists on the Continent.

The rise of the laboratory

Nineteenth-century scientists typically divided into two groups – the natural philosophers, whose interests were either theoretical or laboratory-based, and the natural historians, who were basically concerned with observation. Natural historians in the nineteenth century rarely required much more than a microscope, so their need for a laboratory was limited. Darwin's children were so accustomed to him examining barnacles in his study that they assumed everyone used their studies for this purpose. When Darwin did carry out experiments, he used either his greenhouse, or his garden. The needs of chemists and physicists were different: for them, the nineteenth century saw the development of the modern laboratory.

It is said that an eminent scientist of the early nineteenth century, William Wollaston, was once asked about his laboratory. He summoned a servant, who brought in a tray containing all the apparatus that Wollaston used. Such simplicity could not last for long. Chemists, in particular, needed a special room where they could store their array of chemicals and apparatus, and which was equipped with both water and heating. Most early chemical researchers set aside a room for their studies. Even late in the nineteenth century, a freelance chemist, like Crookes, did much of his work in a small laboratory set up at home. By the second decade of the century, private investigators could obtain ideas on what was needed in a laboratory from readily available texts, such as Parkes' *Chemical catechism*.

The most elaborate example of a laboratory at that time was at the Royal Institution. The Institution had been set up as a private venture at the end of the eighteenth century to act as a focus for the encouragement of applied science. The Managers of the Institution quickly developed its laboratory facilities, and built a lecture theatre next to the laboratory, so that experiments devised there could be displayed to the audiences who attended the lectures. Humphry Davy was appointed as resident chemist in 1802, and attracted large crowds to his lectures. They commented admiringly on his well-devised experiments. The range of available apparatus was soon greater than that to be found in any other contemporary laboratory. In particular, both Davy and Faraday required massive batteries (only invented at the beginning of the century) for generating large amounts of electricity as part of their work on electrochemistry.

In terms of research, the Royal Institution laboratory was intended primarily for a few resident staff. A new development in the nineteenth century was the creation of communal laboratories, where teaching and research could go on together. The idea originated in Germany, more specifically under Justus von Liebig when he was professor of chemistry at Giessen. Since a number of British chemists spent some time doing chemical research in Germany, the approach soon began to appear in Britain. One of the earliest such laboratories was set up in Manchester in the 1840s by Playfair, who had studied with Liebig. He was offered space in some cellars, and learnt the hard way that laboratory design has its special problems.

I had one afternoon three or four organic analyses in operation, the tubes being heated with charcoal. I felt ill and out of sorts, and went to my lodging. Soon afterwards a cab came to take me back to the laboratory. I found to my dismay two of the pupils lying insensible in the area outside, and at once saw they had been poisoned by the fumes of charcoal, as indeed I also had been.... I instantly dashed a pail of water over each of my prostrate pupils, and to my joy found that they revived. This was a practical lesson in ventilation which I never forgot.[4]

The most important laboratory founded prior to the mid-century was the Royal College of Chemistry in London. The layout of the building

A design for a chemical laboratory, as envisaged in the early decades of the nineteenth century. Later in the century much of the equipment on the left of the picture would be replaced by smaller items. From S. Parkes, *The chemical catechism* (Baldwin, Cradock and Joy, 1822).

that housed it was devised by the first professor, Hofmann, who had previously been Liebig's assistant in Giessen. The basement was used for large apparatus, such as furnaces, and for storage. The ground floor was entirely devoted to student laboratories. The floor above housed Hofmann's private laboratory, along with a lecture theatre and a room for delicate activities, such as weighing chemicals. Many of the leading British chemists visited or worked at the College, and it provided them with an example of what a chemical laboratory should look like. It was also influential in showing the type of chemical training that was required – for example, both qualitative and quantitative analysis. Frankland, who succeeded Hofmann, was particularly keen on standardising the teaching of practical chemistry, and subsequently wrote an influential booklet on *How to teach chemistry*.

Prior to this, in the early 1850s, Frankland had been professor of chemistry at the newly created Owens College in Manchester. He and the architect were charged with designing a chemical laboratory. They visited laboratories elsewhere, and Frankland claimed that their resulting design was, 'at that time, superior in convenience for elementary and advanced students, and in lighting, warming and ventilation, to any other laboratory in Great Britain'.[5] The problem was that the demands on laboratories increased throughout the century, as chemistry itself developed. In the 1870s, Henry Roscoe, who, like Frankland, had studied in Germany, helped design new laboratories for Owens College. The layout adopted was widely copied in Britain, and even by Liebig's successor in Germany, so bringing the wheel full circle.

As with chemistry, so with physics. Both private and institutional laboratories appeared in the nineteenth century. Rayleigh's initial experimenting at home was often done on the top of a grand piano. The room in which the piano was situated was also used for family prayers, so Rayleigh had to clear away his equipment every day. After his marriage in 1871, Rayleigh decided to set up a proper laboratory. He converted a stable, dividing it into four rooms. Downstairs contained a workshop and a darkroom for photography. Upstairs there were two rooms: one of them darkened for optical experiments. (Rayleigh could not find a decent matt black paint, so the estate bricklayer developed one based on mixing lampblack with beer.) Rayleigh called on Maxwell for advice on how to equip the laboratory, but developed much of the apparatus himself. For example, he needed to turn off an electric current rapidly, so he set up one of his pistols to shoot through the conducting wire. (A young visitor wrote to her parents that she had not seen Lord Rayleigh when she arrived because he was busy shooting in his lavatory.)

Maxwell's advice was based partly on Kelvin's experiences in setting up a physics laboratory. When Kelvin became Professor of Natural Philosophy at Glasgow, he required laboratory space both for his own research and for teaching students. Around 1850, he acquired a disused wine-cellar in the basement for this purpose. William Ayrton, later a pioneer electrical engineer, was trained there in the 1860s, and recalled what it was like:

[The] laboratory consisted of one room and the adjoining coal-cellar . . . There was no special apparatus for students' use in the laboratory, no contrivances such as would today be found in any polytechnic, no laboratory course, no special hours for students to attend, no assistants to advise or explain, no marks given for laboratory work, no workshop, and even no fee to be paid. . . . students experimented in the one room and the adjoining coal-cellar, in spite of the atmosphere of coal dust, which settled on everything, produced by a boy coming periodically to shovel up coal for the fires.[6]

The emphasis on cellars for chemistry and physics laboratories was often because these provided the only free space available. But also, in the first half of the century at least, chemists needed to use bulky equipment, such as furnaces, while physicists wanted a place as free from vibration as possible. The new laboratories that appeared in the latter half of the nineteenth century were properly designed (usually after discussion between the architect and the professor) to cater for the changing needs of teaching, as well as research. In Kelvin's laboratory, the practical work was basically an extension of the professor's research. As the number of students requiring practical training increased, so more standardised exercises were developed, and the teaching and research activities began to separate. Information on practical work in physics was limited. One of the main sources was a book by Edward Pickering, a professor at the Massachusetts Institute of Technology in the USA. His *Elements of physical manipulation* appeared in 1873, and laid down in some detail how to proceed:

Each experiment is assigned to a table on which the necessary apparatus is kept and where it is always used. A board called an indicator is hung on the wall of the room, and carries two sets of cards opposite to each other, the one bearing the names of the experiments, the other those of the students. When the class enters the laboratory each member goes to the indicator, sees what experiment is assigned to him, then to the proper table where he finds the instruments required, and with the aid of the book performs the experiments.[7]

This method was taken up in what was to become the most famous physics laboratory in Britain – the Cavendish Laboratory. Towards the end of the 1860s, Cambridge examined the possibility of founding a new physics laboratory alongside a new professorship of experimental physics. The problem was finding the money. In 1870, the Chancellor of the University offered to pay for the proposed laboratory; the planning went ahead, and the laboratory was completed in 1874 at a cost of £8450 (perhaps half a million pounds in modern currency). The Chancellor at that time was William Cavendish, Seventh Duke of Devonshire. The Cavendish family was already famous in science due to the experimental investigations of Henry Cavendish in the latter part of the eighteenth century. The seventh Duke was also a well-known figure in the world of science. He had been elected the first President of the Iron and Steel Institute in 1869, and was in charge of a Royal Commission which investigated teaching and research in science during the period when the Cavendish Laboratory was being built. More to the point, he had been Second Wrangler and First Smith's Prizeman when a student at Cambridge: exactly the same record as Maxwell, who was appointed to the new professorship. Maxwell was soon involved in the problems of overseeing the construction of the laboratory.

But at present I am all day at the Laboratory, which is emerging from chaos, but is not yet cleared of gas-men, who are the laziest and most permanent of all the gods who have been hatched under heaven.[8]

At least Maxwell was rewarded with a specially designed building. The new university colleges that sprang up round the country during the latter decades of the nineteenth century were typically housed in existing buildings that had to be converted as best as possible. In the early 1880s, a former lunatic asylum in Liverpool was converted to house the new university college there. Lodge went up to supervise the construction of the laboratories.

The padded room had not been pulled down when I went there, but it disappeared along with the rest, and became part of my laboratory. These makeshift laboratories, thus thrown together in a building intended for other purposes, have both advantages and disadvantages. Their disadvantages are obvious, but their advantages are not to be despised. There is nothing sacred about the walls that are left; they can be plugged and pulled about, and holes cut in them wherever wanted. By ingenious scheming, fittings can be adapted to the circumstances, and where these are successful you can be pleased

The physicist Oliver Lodge after he had become the Principal of the new University of Birmingham. From *Vanity Fair* (4 February 1904).
Newspaper Library, Colindale

accordingly; while, if any are unsuccessful, you have nothing to blame yourself for.[9]

Standardising practical work

Agreement on a standard approach to both laboratory design and practical work in chemistry and physics was reached in the few decades after the middle of the nineteenth century. The approaches developed then proved so successful that they dominated teaching in these two subjects for the next hundred years, and still provide the framework for practical teaching today. Nor was this development limited to the physical sciences. Organised laboratory training in biology also appeared at this time, and similarly influenced practical work far into the future. The idea of using dissection for teaching in botany and zoology already existed by 1870, as did student use of microscopes for examining tissues. But it was Huxley and his colleagues who integrated these into a systematic form of biological training. As a contemporary commented:

Botanists had always been in the habit of distributing flowers to their students, which they could dissect or not as they chose; animal histology was taught in many colleges under the name of practical physiology; and at Oxford an excellent system of zoological work had been established by the late Professor Rolleston. But the biological laboratory, as it is now understood, may be said to date from about 1870, when Huxley, with the cooperation of . . . others, held short summer classes for science teachers at South Kensington, the daily work consisting of an hour's lecture followed by four hours' laboratory work, in which the students verified for themselves facts which they had hitherto heard about and taught to their unfortunate pupils from books alone. The naive astonishment and delight of the more intelligent among them was sometimes almost pathetic. One clergyman, who had for years conducted classes in physiology under the Science and Art Department, was shown a drop of his own blood under the microscope. 'Dear me!' he exclaimed, 'it's just like the picture in Huxley's *Physiology*'.[10]

Needless to say, the growth of widespread practical teaching led to a greater attention to safety requirements. For example, fume cupboards for handling unpleasant or dangerous materials became commoner. The very fact that practical teaching was becoming more standardised helped. When J. J. Thomson took the physics course at Manchester in the early 1870s, Balfour Stewart, the professor there, still ran the old system of laboratory teaching. Students could work on what they liked,

J. J. Thomson in his days as a physicist at the Cavendish Laboratory, sporting a
well-cultivated moustache. From W. A. Tilden, *Chemical discovery and invention
in the twentieth century* (George Routledge, 1917).
The British Library, 8903.aaa.19

often setting up an experiment that interested their professor. Thomson recalled:

Stewart was trying to find out whether there is any change in weight when substances combine chemically. . . . He asked me to make the weighings. I had made a good many without finding any difference or without there being any explosion of any kind. One Saturday afternoon, however, when I was alone in the laboratory, after tilting the flask, though the mercury ran over the iodine no combination took place. I held it up before my face to see what was the matter, when it suddenly exploded; the hot compound of mercury and iodine went over my face and pieces of glass flew into my eyes. I managed to get out of the laboratory and found a porter, who summoned a doctor. For some days it was doubtful whether I should recover my sight. Mercifully I did so, and was able after a few weeks to get to work again.[11]

New methods of training made student practicals somewhat less exciting, but the original research carried out by their mentors continued to have its hazards. Roscoe, also at Manchester, recorded one of his own research experiences:

. . . as I was filtering a few cubic centimetres of the liquid into a test-tube, the whole thing exploded, the bottom of the test-tube bored a hole, an inch in diameter, almost through the hard wood of the filter-stand, whilst the glass was shivered into many thousand fragments in my left hand, from which I afterwards picked out some 200 pieces.[12]

The main problem for researchers occurred when a hazard went unsuspected. Mercury, for example, was a common chemical in the laboratory: Faraday made extensive use of it. But it took a long time before the effects of mercury poisoning were recognised. At the end of the century, researchers worked with the newly discovered radioactive materials and with X-rays without taking any precautions. Many of Crookes' surviving laboratory notebooks are still radioactive today. Earlier in the nineteenth century, the adverse effects of ultraviolet light were not recognised immediately. Stokes recorded a visit to Lockyer in South Kensington:

I found him in his Laboratory with one ear, and the part of the neck near it, as red as a turkey cock. He told me that on Monday he had been working with the electric arc given by a powerful Siemen's machine. Knowing the injurious effect of the light upon the eyes, he took care not to look at it, so he stood with his back to the light about 2 feet off. Yet it affected the skin of the back of his neck which was exposed to it . . . He said that the day before the skin had been black, and that wanting to write something he was obliged to dictate it, as the inflammation had extended to his eyes.[13]

Equipment and technical assistance

The making of scientific instruments was well established in Britain by the nineteenth century. For example, sophisticated surveying, navigational and astronomical equipment was available from a number of makers, and some were already constructing equipment that could be used in chemistry – thermometers, air pumps, balances, and so on. The development of new research techniques in the nineteenth century required the invention and construction of new types of instrumentation. For example, the growing interest in chemical analysis led the Royal Institution to introduce that symbol of modern chemistry – the test-tube. A more important advance was the gas burner designed in Bunsen's laboratory in Germany during the 1850s. Previously, chemists had had to rely on bulky furnaces for their heat: now their 'Bunsen burners' provided individual heat sources on the bench-top. This innovation illustrates the role of infrastructure in the development of new techniques. Bunsen burners were easy to install because they used gas from the gas mains that began to be built in the first half of the nineteenth century.

In response to these changes, some of the existing instrument-makers expanded their repertoire, but a number of new names also appeared. An interesting example is John Griffin, who had attended chemistry lectures in Glasgow in the early 1820s. The lectures were illustrated by experiments, though no hands-on facilities for students were available. So Griffin wrote a book to encourage experimentation at home. The reasons he gave lay behind much practical teaching throughout the century.

The hearing of lectures, and the reading of books, will never benefit him who attends to nothing else; for Chemistry can only be studied to advantage practically. One experiment, well-conducted, and carefully observed by the student, from first to last, will afford more knowledge than the mere perusal of a whole volume. It may be added to this, that chemical operations are, in general, the most interesting that could possibly be devised – Reader! what more is requisite to induce you to MAKE EXPERIMENTS?.[14]

Griffin, whose family background was in bookselling, continued to write on chemistry, and in the latter part of the 1830s began to combine this with selling the equipment that his books described. In 1841, he produced a first catalogue of the chemical apparatus he could supply. At the end of that decade, he decided to move to London, where his products were increasingly designed to fit in with the growing market for practical chemistry teaching at colleges and schools. His firm became a major player in this field in the latter half of the nineteenth century, and his catalogues reflect the increasing variety of apparatus that could be bought off the peg. There is an evident contrast here with the situation earlier in the century as portrayed in Faraday's *Chemical manipulation*, published in 1827. Faraday assumed that the chemist would have to make almost all the apparatus himself, using bits and pieces that came to hand. In terms of the chemicals that chemists used, such self-help remained the standard approach until towards the end of the century. Chemicals bought from manufacturers always contained some impurities that might affect analyses, so chemical researchers had to learn how to purify reagents for themselves.

The development of laboratory training, and of laboratory-based research, in the latter part of the nineteenth century meant that the apparatus required became more diverse, and more money had to be spent on it. For example, the Duke of Devonshire initially contributed the funds needed to equip the Cavendish Laboratory with the apparatus necessary for its work. Within a few years, Maxwell was using several hundred pounds of his own money to update and expand the amount of equipment available. His successor, Lord Rayleigh, found that he had to do the same, using £500 of his own money plus another £1,000 donated by friends. Even so, as Maxwell observed, there was a problem:

It has been felt that experimental investigations were carried on at a disadvantage in Cambridge because the apparatus had to be constructed in London. The experimenter had only occasional opportunities of seeing the instrument maker, and was perhaps not fully acquainted with the resources of the workshop, so that his instructions were imperfectly understood by the workman. On the other hand, the workman had no opportunity of seeing the apparatus at work, so that any improvements in construction which his practical skill might suggest were either lost or misdirected. During the present term a skilled workman has been employed in the laboratory, and has already greatly improved the efficiency of several pieces of apparatus.[15]

The value of having technicians attached to a laboratory, as well as having instrument-makers close to hand, had been recognised before this. Kelvin actually turned down the offer of the professorship at Cambridge: one of his reasons was that the University could not provide the same resources that he had in Glasgow. In due course, however, Cambridge was able to compete on more equal terms. Darwin's son, Horace, was very interested in scientific instrumentation. In the 1870s,

Cambridge practical teaching was developing rapidly – not only in physics and chemistry, but also in other areas such as physiology and engineering. Horace set himself up there as a maker of specialised apparatus. In due course, this led to the establishment of the Cambridge Scientific Instrument Company, often referred to in the Darwin family and in Cambridge as 'The Shop'. It was soon sufficiently well-known in the University that the introduction to its catalogue was used as an exercise for Greek translation. So far as the Cavendish Laboratory was concerned, Horace Darwin was quickly involved with Rayleigh, and with his successor, J. J. Thomson, in providing and repairing specialised laboratory equipment.

Successive professors at the Cavendish managed to appoint skilled craftsmen to look after the laboratory workshop. Indeed, the superintendent in Rayleigh's time left with him to help with the experiments that Rayleigh intended to carry out at home. One of Thomson's chief technicians was the son of Horace Darwin's foreman. The two men – Pye senior and junior – resigned from their respective jobs at the end of the century in order to set up what was to become one of the main radio manufacturers in the UK. Senior technicians, like these two, might earn a shilling an hour, and often helped with the laboratory teaching. This included passing on such skills as glass-blowing. For, despite the increased availability of manufactured apparatus, students – and especially research students – were still expected to improvise as necessary.

Well-equipped workshops only became an established feature in the latter part of the nineteenth century. Before that, individual researchers who needed special equipment typically organised things for themselves. For example, Babbage spent many years building his computer. This required a range of skills, so he set up his own system:

In order to carry out my pursuits successfully, I had purchased a house with above a quarter of an acre of ground in a very quiet locality. My coach-house was now converted into a forge and a foundry, whilst my stables were transformed into a workshop. I built other extensive workshops myself, and had a fireproof building for my drawings and draftsmen. Having myself worked with a variety of tools, and having studied the art of constructing each of them, I at length laid it down as a principle – that, except in rare cases, I would never do anything myself if I could hire another person who could do it for me.[16]

Babbage's problem was that some of what he wanted was beyond the capabilities of contemporary instrument makers. The same was true of

William Parsons, Third Earl of Rosse. Inspired by the large telescopes that William Herschel had built, Rosse decided to make the largest telescope ever. He built the telescope on a site at his home in Parsonstown in Ireland. Its construction entailed erecting a fully equipped workshop and then finding the workmen to use it. Rosse was much involved in the manual operations himself, working alongside his employees. He encouraged his sons to join in workshop activities: the most enthusiastic of them, Charles Parsons, later became one of the leading marine engineers in Britain. The telescope, when completed, was, indeed, the largest in the world: it was popularly known as the 'Leviathan of Parsonstown'. One of Rosse's friends celebrated by walking through the tube of the telescope wearing a top hat and with a raised umbrella over his head.

The main difficulty with the new telescope was moving it to follow the stars with the necessary accuracy. Problems of this kind, along with the increasing sophistication of the telescopes themselves, meant that by the latter part of the nineteenth century, individuals were effectively restricted to the construction of relatively small telescopes. Correspondingly, firms such as Grubb in Dublin – with which Charles Parsons was later involved – were created to cater for those who required something more. Howard Grubb, speaking at the Royal Institution in the 1880s, emphasised just how much time and training were needed to carry out such work:

A well-known and experienced amateur in this work declared his conviction that no one could learn the process under nine years' hard work, and I am inclined to think that his estimate was not an exaggerated one.[17]

Besides technicians, researchers often engaged assistants to help with their work. This was particularly common when the researcher had another job that required his attention. Faraday effectively started life at the Royal Institution as a technician and graduated to the role of assistant when he demonstrated his research capabilities. Later in the century, assistants were typically well-educated and often moved on to more important posts after their period of assisting. McLeod and Pedler, who assisted Frankland in the 1860s, both subsequently became Fellows of the Royal Society. So did Lockyer's research assistants, Meldola and Fowler. Another of Lockyer's assistants, Gregory, followed him as Editor of *Nature*. One of Lockyer's sons, Jim, also acted as his assistant for many years, Keeping things in the family was not uncommon in the nineteenth century. Thus, John Herschel started his research career by helping his father. In universities and colleges, assistants, besides their research activities, were often used as demonstrators in the laboratory. At the Cavendish Laboratory, for example, the practical teaching was looked after for many years by William Shaw and Richard Glazebrook. Shaw later became Director of the Meteorological Office, while Glazebrook became Director of the National Physical Laboratory. At Oxford and Cambridge, such people could seek the security of a college fellowship. At the new colleges and universities of the late nineteenth century, assistants usually occupied an ill-defined position, with low salaries and often no security of tenure. It was not until well into the twentieth century that this position changed.

In terms of his initiation into practical research, Darwin had much in common with other naturalists in the first half of the nineteenth century. They saw travel as a way of increasing their knowledge, and accepted that, though field trips beforehand could help them, yet they must often develop research skills on their own. For researchers in chemistry and physics, the situation was rather different. They needed equipment and the space to use it. For a wealthy man, this might present no problem, but he would still need some training in experimental techniques. For both the wealthy and the less wealthy, the communal laboratory facilities offered by some institutions could help in this respect. By the middle of the nineteenth century, questions concerning the design and organisation of such laboratories were beginning to be asked. In the latter half of the century, training in laboratory methods became widely available. Even biological training now became systematised, and increasingly laboratory-based: significantly different from the traditional approach to training naturalists. Though self-taught researchers continued to appear – Lockyer, for example, in the 1860s – the trend in science was clearly away from the image of the self-sufficient individual. In this respect, there is a clear difference between Lockyer, who used sophisticated equipment and employed able research assistants, and Darwin, who did not.

Scientific links and communication

Families and friends

THE nineteenth century was notable for its business dynasties, with son following father in the family firm. Warren De la Rue, a pioneer of astronomical photography, spent much of his time in the family printing firm – still a famous name in the business world today. Indeed, he invented the first envelope-making machine, and exhibited it at the Great Exhibition in 1851. But along with such business families, the nineteenth century had an equally notable number of scientific dynasties. The obvious example is the Darwin family. Of Darwin's two famous scientific forefathers, Erasmus Darwin just survived into the new century, while Charles' other grandfather, Josiah Wedgwood – founder of a business dynasty of his own – died in the 1790s. Darwin's father was a Fellow of the Royal Society, as was Charles, and as were three of his sons – George, Francis and Horace. More distant relatives also had scientific links. Josiah Wedgwood's son, Thomas, was a close friend of Humphry Davy, and was, himself, a pioneer of photography. Francis Galton, whose name crops up in several different areas of nineteenth-century science, was Charles' half-cousin, and Galton's grandfather had been a member of the Lunar Society along with Erasmus and Josiah. The Darwin family was not alone in terms of scientific eminence. William Herschel, though a contemporary of Erasmus Darwin's, survived longer, into the 1820s. He was a Fellow of the Royal Society, as was his son, John, as were two of John's sons, another John and Alexander. William's sister, Caroline, was the only woman to be awarded a gold medal of the Royal Astronomical Society in the nineteenth century.

In some cases, families had links with particular places or institutions. For much of the nineteenth century, Kew Gardens was ruled by a family dynasty. William Hooker was succeeded as Director by his son, Joseph, who was, in turn, succeeded by his son-in-law, William Thiselton-Dyer. Joseph married the eldest daughter of Darwin's

The polymath Francis Galton at work on one of his innumerable projects. From his book,
Memories of my life (Methuen & Co., 1908).
The British Library, 010827.ee.14

botanical mentor at Cambridge, Henslow. Again, Kelvin's father was a professor of mathematics first at Belfast and then at Glasgow. Kelvin's brother, James, was subsequently a professor of engineering at the same two places, while Kelvin, himself, was a professor at Glasgow. Robert Ball's family concentrated on Dublin. He, himself, was professor of applied mathematics and then Astronomer Royal for Ireland there. One brother was professor of geology and another professor of surgery in the same city. As the Hooker dynasty indicates, a number of nineteenth-century scientists were related to each other by marriage. Perhaps the most extended example relates to three eminent chemists, William Perkin, Frederic Kipping and Arthur Lapworth. All became Fellows of the Royal Society, and were linked by their work – Kipping had assisted Perkin, and Lapworth had assisted Kipping. They then became linked as brothers-in-law, marrying three sisters, the Misses Holland.

Because the number of scientists was small, even those without scientific connections in the family soon met others in the scientific world. Before Faraday joined the Royal Institution, he became a member of a discussion group called the City Philosophical Society. One of the other members, whom he came to know well, was Richard Phillips, a chemist who became editor of the *Philosophical Magazine*. This was an influential scientific journal in which Faraday was later to publish much of his work on electricity. Such links via clubs or institutions were common. Perhaps the most enduring connections were those set up during training. For example, scientists who had been to Cambridge, especially those who had taken the Mathematics Tripos, often remained in touch during their subsequent careers. This was true of John Herschel, Babbage and Peacock – the trio who helped change the mathematics teaching at Cambridge in the 1820s – though each subsequently went their own way, with Peacock becoming Dean of Ely Cathedral. This network of communication sometimes became a fully-fledged 'old boys' network. For example, when Airy became Astronomer Royal at Greenwich in the 1830s, he appointed another Cambridge man to be his chief assistant. The post continued to be filled by Cambridge men for the next hundred years.

Along with this Cambridge network, there was another grouping, equally influential, in London. Unlike Cambridge, it involved a number of institutions. In the first half of the century, the Royal Institution was the main focus; in the latter half, the group of institutions at South Kensington took centre stage. University College and King's College

Victorian scientists showed great ingenuity in arranging their feasts. This one was held inside a model of an Iguanodon. (Note that Victorian reconstructions of dinosaurs do not always agree with modern reconstructions.) The names of the leading experts on animal fossils are attached to the roof of the tent. From *Illustrated London News* (7 January 1854). The British Library. PP7611

played supporting roles. It was said that, while William Ramsay was professor at University College, no eminent foreign chemist visited London without becoming his guest. Though Cambridge continued to dominate in mathematics and physics, London took the lead in chemistry and biology. The growing London influence can be seen in the importance that came to be attached to the informal X-Club, which appeared there in the mid-1860s. As Herbert Spencer explained, somewhat pompously:

In pursuance of a long-suspended intention, a few of the most advanced men of science have united to form a small club to dine together occasionally. It consists of Huxley, Tyndall, Hooker, Lubbock, Frankland, Busk, Hirst and myself. Two more will possibly be admitted. But the number will be limited to ten. Our first dinner was on last Thursday; and the first Thursday of every month will be the day for subsequent meetings.[1]

Some of the members of this group, along with their friends, earned the nickname of the 'Young Guard of Science', because they were seen as reacting against the previous generation of scientists (the Old Guard). X-Club discussions were typically relaxed: Huxley recalled talking about 'politics, scandal, and the three classes of witnesses – liars, damned liars, and experts'. But because of the growing eminence of the members – who produced three successive Presidents of the Royal Society – their discussions could, on occasion, affect the wider scientific community. Huxley records one exchange of conversation that reflects the external view of the Club:

. . . two distinguished scientific colleagues of mine once carried on a conversation (which I gravely ignored) across me, in the smoking-room of the Athenaeum, to this effect, 'I say, A. do you know anything about the X-Club?' 'Oh yes, B., I have heard of it.' 'What do they do?' 'Well, they govern scientific affairs, and really, on the whole, they don't do it badly'.[2]

Societies

It used to be said that, if three Englishmen were stranded together on a desert island, they would immediately form a club or society with one of them as president, another as secretary, and the third as treasurer. (A distinction was often drawn in the nineteenth century between 'societies', which were supposed to deal with knowledge, and 'clubs' which were meant for socialising, though the two frequently meshed together.) Certainly, science in the British Isles during the nineteenth century seems to fit this picture. By the 1840s, London hosted an array of specialist scientific societies, most of them set up since the beginning of the century – for example, for geology (founded 1807), astronomy (1820), zoology (1826), botany (1836), chemistry (1841). Life in fashionable London concentrated on the period from November to July, when the Houses of Parliament were sitting. Most learned societies fixed their meetings for these months (and publishers tried to produce new titles then). Faraday believed there were four ways of gathering information – by conversation, by lectures, by reading and by observation or experiment. He thought that membership of a society could help with all of these. But all were agreed that socialising with fellow-scientists was an equally important activity.

The Royal Society had a dining club to which a limited number of Fellows were admitted. The new specialist societies followed the same pattern. As Archibald Geikie, a leading geologist in the latter years of the century, recalled, such dining clubs were, 'gatherings of the more prominent members to dine and talk and thereafter to adjourn to the evening meeting of the Society. Besides promoting good-fellowship among the members, they gave opportunities for much pleasant scientific gossip'.[3] The British Association for the Advancement of Science was set up in the 1830s as a way of bringing science to the provinces. Though it moved to a new venue each year, it still had a dining club – the Red Lions. The name, in this case, came from the tavern where the club met for the first time. Taverns were popular meeting places. One in London – the Freemason's Tavern – actually gave birth to two of the new scientific societies. As a founder of the Geological Society recorded in his diary for 1807: 'dined at the Freemason's Tavern, about five o'clock, with Davy, Dr. Babington, etc, etc, about eleven in all. Instituted a Geological Society'.[4] Some years later, John Herschel recorded in his diary for 1820: 'Dine at the Freemason's Tavern . . . to consider of forming an Astronomical Society'.[5]

The emergence of these societies reflected the growing interest in specific areas of science, but this was far from being their whole motivation. Another was the limitations of the Royal Society. Under the long reign of Banks as President, its pre-eminence as a scientific body had become increasingly questioned. In the 1820s, John Herschel complained that there was no point in trying to compete with, 'our Continental neighbours whether French or German in matters of science generally. Our day is fast going by'.[6] Herschel believed that the

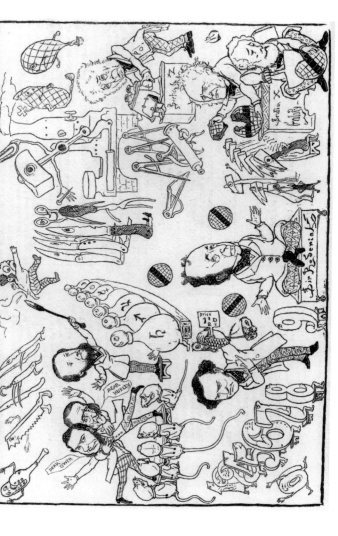

A cartoon showing some of the main participants at the British Association meeting of 1865. Murchison, at that time President of the Royal Geographical Society, is seated at bottom centre, juggling with terrestrial globes. The next figure clockwise from him is Babbage, lecturing to numbers. Then come the embracing couple of Huxley and Owen: their controversy over evolutionary ideas was currently at its height. To the right of them is Tyndall: the hot poker he is holding refers to his experiments on heat. On the far right is the astronomer John Herschel, with the geologist Lyell below him. Finally, the figure in the bottom right-hand corner is another well-known geologist of the time, George Scrope. From *Punch* (23 September 1865).

The British Library, PP5270

Royal Society was to blame for failing to give effective leadership. His friend Babbage agreed with him, and not long afterwards published a book entitled *Reflections on the decline of science in England*. This called for changes that would, 'rescue the Royal Society from contempt in our own country, from ridicule in others'.[7] Though the attack by Babbage, Herschel and others seemed to achieve little at the time, yet it led ultimately to the reorganisation of the Royal Society's activities. In the latter part of the 1840s, a reform of the Society's statutes was agreed, paving the way to a more professional society. As a result, the Society came to play an important role in the development of science in Britain during the latter half of the century.

Meanwhile, not everything was sweetness and light in the new societies. Darwin, on his return from his *Beagle* voyage, met up again with fellow zoologists. He soon decided: 'I am out of patience with the zoologists, not because they are overworked, but for their mean, quarrelsome spirit. I went the other evening to the Zoological Society, where the speakers were snarling at each other in a manner anything but like that of gentlemen'.[8] Later in the century, Lockyer and his friends resigned from the Council of the Royal Astronomical Society because of a feud with the Secretary and his friends. Lockyer explained to Airy:

Permit me then to tell you at once in confidence that my primary reason for quitting the Council is the offensive manner in which Mr. Proctor is conducting himself towards me. Week after week in more or less obscure journals which as Editor of *Nature* I must see I find myself attacked by one who takes good care to advertise himself as 'Honorary Secretary of the Royal Astronomical Society'. Now to Mr. Proctor as an individual I do not care to reply. To deal with him as the Honorary Secretary of an honourable Society would cause a scandal and on these grounds I have determined not to reply to him.[9]

Travel and communication

The Royal Society, though centred on London, included members from elsewhere in the country. Until the nineteenth century, limitations on travel made it difficult for more distant members to play much part in the Society's activities. In the eighteenth century, local scientific groups, such as the Lunar Society in the Birmingham area, might provide easier contact. It was convenient to have a local meeting place which could be reached on horseback. Indeed, the Lunar Society's name derived from the fact that members met on the Mondays closest to full moon, so that their horses could find the way home more easily. By the early nineteenth century, stage-coaches had greatly speeded up travel times. In the mid-eighteenth century, the trip from Edinburgh to London took ten days; by the 1830s, this had been cut to two days. Cross-country journeys were less well-served.

This speeding up of traffic allowed people from further away to attend meetings in London. The new subject-based societies were based in London, as was often indicated in their full title – for example, the Geological Society of London, or the Zoological Society of London. But they were increasingly expected to function as national foci of interest in their subjects. The advent of railways in the 1830s further enhanced the importance of societies in the metropolis, since all the major railway companies in England converged on the capital. Railway travel had the advantage of being faster, cheaper and more comfortable than travel on stage-coaches. As the London-based societies increased in importance, so the big provincial societies began to decline. The Manchester Literary and Philosophical Society, for example, which had achieved national significance in Dalton's day, gradually lost that position as the century progressed. Dalton's pupil, Joule, was a leading light of the Manchester society. But he recognised the need to become involved in scientific activities in London:

I am going to attend the first meeting of the Council of the Royal Society on Thursday next. When it was proposed I, at once, expressed my willingness to serve on the principle that country Fellows should take their share in the business of the Society. Otherwise it would fall into the hands of too few to give personal attention. I am afraid I shall not be able to attend as many meetings as I ought. One thing is in my favour, i.e. the competition between the Gt. Northern and L. & N.W. Rwys whereby I can go to London on Wednesday and return on Sat. for the double journey 1st class £1.1/-.[10]

Greater ease of attending meetings was backed by a greater ease of communication in general. In the early decades of the century, sending mail was often a slow and expensive process. Letters were typically restricted to a single sheet and cost of delivery depended on distance. All this changed in the 1840s, after Rowland Hill introduced a standard postage rate regardless of distance and railways became important carriers of the mails. In London, where postal deliveries were frequent, it was possible to send a note in the morning, receive a reply in the afternoon, and send a further note in the evening. Scientists who were away from their laboratory could keep track of progress. Raphael

Meldola was one of the leading chemists in the latter years of the nineteenth century. He corresponded extensively with his assistants. It is said that they were quite likely to get a suggestion for an experiment by one post, followed by a card cancelling it a few hours later. Sending printed matter could be expensive, and members of societies were often expected to collect copies of the society's publications from its office to save postal charges. Post cards could be sent for half the cost of a letter, so frequent correspondents often corresponded in this way. The main problem was the lack of space for writing. Maxwell, Kelvin and Tait, who were constantly in touch, developed a shorthand of their own to cram in the maximum information.

Rowland Hill was an inventor, rather than a scientist, though he was elected a Fellow of the Royal Society. The next speeding up of communication was more dependent on the scientific community. This was the development of the electric telegraph in the 1830s. The initiative here was taken by Charles Wheatstone, a professor at King's College London, working with his business partner, William Cooke. The new system of telegraphs was first used extensively for communication by the rapidly expanding railway network, but scientists were involved from the start. The obvious example is the Royal Observatory at Greenwich. By the 1850s, Airy was providing time signals from Greenwich for distribution by the railway telegraphs. (This was how 'Greenwich time' began to dominate as the national standard.) He characteristically observed: 'I cannot but feel a satisfaction in thinking that the Royal Observatory is thus quietly contributing to the punctuality of business through a large portion of this busy country'.[11]

The telegraph system became particularly important for scientists when international links were set up, for it allowed them to communicate with overseas colleagues. Such links entailed the development of submarine telegraphy. Wheatstone was again involved in this, but the main scientific contribution came from Kelvin, who devised a method whereby the weak signals sent via a submarine cable could be readily detected at either end. Kelvin was greatly involved in the whole process of setting up a viable transatlantic cable in the 1860s. His work required frequent commuting between Glasgow and London, and was regarded as so important that the night mail train was held for him at Glasgow on a number of occasions while he completed his experiments. Kelvin's knowledge of scientific instrumentation led to an invitation to act as a judge at an exhibition in the USA in 1876. One of

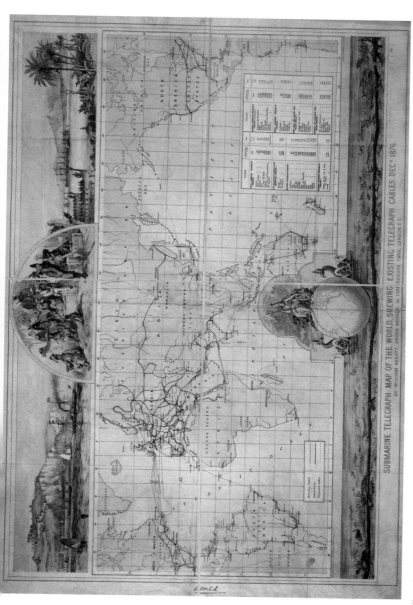

Victorian scientists made a considerable contribution to the development of the submarine telegraph network shown in this map from the 1870s.

The British Library, 957 (14)

the exhibits there was the new telephone invented by Graham Bell. Kelvin brought two of Bell's instruments back with him, and demonstrated the system at the next British Association meeting. Bell subsequently visited Britain, and, by the 1880s, telephone exchanges were springing up across the country. Scientists tended to be early users of the telephone. When, in the mid-1880s, an exchange was set up in Cambridge, the first twenty-five subscribers included George, Francis and Horace Darwin.

Societies and communication

One result of the problems at the Royal Society in the first half of the nineteenth century, and of the strong bias towards London as a place for holding scientific meetings, was the creation in 1831 of the British Association for the Advancement of Science. An association had been set up in Germany some ten years before to bring scientists from the various German states together. It met annually, each time in a different town, and had been notably successful in publicising the value of scientific research. Calls were made for a similar organisation in Britain, especially by Babbage in England and by Brewster in Scotland. (In the early decades of the nineteenth century, Scotland excelled England not only in its provision of science education, but also in its knowledge of what was happening in the scientific world abroad.) As Murchison, an enthusiastic supporter, explained, the Association was intended to be:

a gathering of men of science to give a more systematic direction to their researches, to gather funds for carrying out analyses and inquiries, to gain strength and influence by union, and to make their voice tell in all those public affairs in which science ought to tell.[12]

The first meeting in York was a somewhat precarious affair, since it was opposed by many in the scientific world who saw London as the appropriate centre for scientific meetings. Yet the Association did more than survive. Its next four successful meetings in the main academic centres outside London – Oxford, Cambridge, Edinburgh and Dublin – gave it considerable credibility as an occasion for diffusing scientific ideas. For example, Whewell, who had initially been suspicious of the idea, soon threw his weight behind the organisation. As Murchison observed, this made a difference, for Whewell, 'is the best fattened hack-hunter that we ever had in our stable, for all work comes alike to

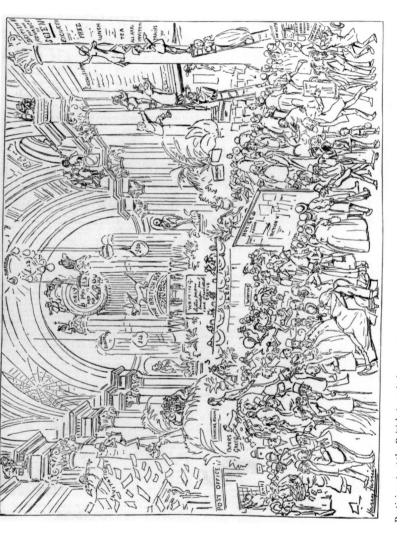

Participants at the British Association meeting in Leeds in 1890. The cartoonist has added the popular abbreviation of its name, 'British Ass', to the organ in the mid-distance, along with a picture of a braying ass. From *Punch* (20 September 1890). The British Library, PP5270

him; no day is too long and no fence too high'.[13] (Murchison's words always reflected his main interests in life.)

In its early years, the Presidents of the British Association were drawn as much from those with aristocratic backgrounds as from amongst the eminent scientists. The same was true for the Royal Society. In both cases, it was seen as important to have an influential figurehead. As Murchison commented regarding the choice of presidents for the British Association: 'we can never again venture to propose a mere man of science except at the great universities. In a mixed great society . . . it is absolutely essential that some public person should be at our head'.[14] This attitude became a cause of concern to members both of the British Association and of the Royal Society as the century progressed, and as the confidence of scientists increased. After the middle of the century, eminence in scientific research became the main criterion when selecting a president for either body. Perhaps the turning point came with the appointment of the Earl of Rosse as President of both the British Association and the Royal Society during the 1840s, for he had dual qualifications, being both an aristocrat and an eminent astronomer.

The avowed purpose of most scientific societies was to assist communication between scientists with similar interests. This was to be achieved especially by holding meetings and publishing journals. The two activities usually went hand-in-hand, with papers that were presented orally at meetings subsequently being published in the society's journal. When the Royal Astronomical Society produced its new journal at the end of the 1820s, the editor explained why it was necessary:

The public is hereby brought more immediately into contact with the Society – the labours of its contributors are canvassed and discussed, while the interest of the author in his subject is yet warm, and when the interchange of ideas respecting it is most beneficial, not only to the public, but to the author himself, whose views may, and probably in many instances will be enlarged or corrected by such intercourse. An authentic, and at the same time public, record is, as it were, opened, of the papers read, and the outlines of their contents rendered matter of history;- thus affording ready means of establishing the claims of authors to priority of discovery.[15]

The emphasis on the 'public' here may seem surprising; but, in the early part of the century, reading and commenting on science was seen as a part of the general culture, much like reading and commenting on poetry. As the century passed and science became increasingly complex, this assumption faded, to an extent which varied with the branch of science. Observational sciences, such as astronomy, geology, meteorology, or natural history, remained popular matters for public discussion throughout the century. Darwin and Hooker, for example, both published notes about their research in the *Gardeners' Chronicle*.

Competition

The quoted editorial has a sting in its tail – priority claims. As the nineteenth century progressed, a scientist's eminence was increasingly weighed in terms of the quality and quantity of the publications he produced. The amount of scientific research increased, and the results had to be disseminated and assessed. Problems arose for the scientific societies, and for the publishing system in general, as they tried to adjust to the new demands this created. William Spottiswoode summarised the pressures in his Presidential Address to the Royal Society in the early 1880s. He was in a good position to assess what was happening. To the scientific world, he was known for his work in mathematics and physics (and as the ninth member of the X-Club). To the wider world, he was known for his involvement in publishing, both via his family involvement in Eyre & Spottiswoode, the Queen's printer, and via his mother, who was the eldest daughter of Thomas Longman, the publisher. His words reflect the common feeling that science had become better organised, but less gentlemanly, during the century:

It may be a matter of regret, although doubtless inevitable, that the same causes that have affected the social, the intellectual, the industrial and the political life of our generation, and have made them other than what they were, should affect also our scientific life; but, as a matter of fact, if science is pursued more generally and more ardently than in former times, its pursuit is attended with more haste, more bustle, and more display than was wont to be the case. Apart from other reasons, the difficulty, already great and always rapidly increasing, of ascertaining what is new in natural science; the liability at any moment of being anticipated by others, constantly present to the minds of those to whom priority is of serious importance; the desire to achieve something striking, either in principle or in mere illustration; all tend to disturb the even flow of scientific research.[16]

What Spottiswoode is regretting, of course, is that science was increasingly being seen as a competitive professional activity. One aspect of this was the changing attitude to exploration of a new line of research. In the earlier years of the century, there had been a tacit

The chemist William Ramsay pointing at the list of noble gases he had helped discover (the elements in Group VIII on his chart). From *Vanity Fair* (2 December 1908). Newspaper Library, Colindale

understanding that a person who lighted on something new would be given a certain amount of time to explore it before others joined in. As Ramsay explained towards the close of the century, 'it is precisely after publication of the original idea, that sufficient time should be allowed to elapse, so as to give the author time to develop his idea'.[17] This attitude was backed by the strong Victorian sense of property, but had been eroded by the time Ramsay was writing. Amusingly enough, Ramsay was widely regarded as a leader of this erosion. In the 1890s, Rayleigh discovered the first of the noble gases, argon, and Ramsay immediately started investigating it, much to the annoyance of Rayleigh's scientific friends. Rayleigh, himself, was more annoyed that Ramsay wanted to publish as quickly as possible before the properties of the new element had been fully explored in the laboratory. Rayleigh and his friends reflected the older approach to science: Ramsay the newer approach.

The underlying problem was that priority in making a scientific advance was (and is) normally given to the person who publishes first. So rapid publication was very desirable. For much of the nineteenth century, scientists were faced with the need to streamline a publishing system designed for a more leisurely age. Most new research was made public via a scientific society. Typically, a paper was presented orally at a meeting of the society, either by the author, or by an officer of the society. For example, the famous joint paper on evolution by Darwin and Wallace was read to the Linnean Society by its Secretary, because Darwin was unwell and Wallace was abroad. By the latter part of the century, the influx of papers in some societies was so large that a number of them might be noted, but not actually read. Even so, delays in publication could be lengthy. While the Rayleigh/Ramsay debate was going on, a zoologist complained bitterly about his own experience:

In the winter of 1894–95, I completed a piece of work . . . and was advised to offer it to the Zoological Society for publication. The paper was received, in the first instance, on June 6, and I hoped it would have been taken as read at a meeting of the Society held in that month. It was, however, not read till November 19 . . . It was ordered for publication in the *Trans.*, and now (November 14, 1896) nearly twelve months from the date of reading, I have not yet received my proofs. Surely such an extraordinary delay as this ought not to be necessary. . . . I have just suffered the chagrin of seeing a paper embodying a large slice of my results published by an Italian journal.[18]

Even setting aside the question of oral presentation at a meeting, a paper had to pass through a number of stages before it appeared in print. Each of these stages introduced a delay. The first was assessment of quality – was the paper worth publishing? This was usually decided by the secretary of the society, though, as the number and complexity of papers grew throughout the century, he was increasingly likely to call on advice from others. One drawback with seeking advice was again the Victorian sense of property. An expert in a particular topic might well dislike the idea of others working in his field. Huxley reported on a paper he had submitted to the Royal Society:

> I know that the paper I have sent in is very original and of some importance, and I am equally sure that if it is referred to the judgement of my 'particular friend' that it will not be published. You will ask with some wonderment, Why? Because for the last twenty years [he] has been regarded as the great authority on these matters, and has had no one to tread on his heels, until at last, I think, he has come to look on the Natural World as his special preserve, and 'no poachers allowed'. So I must manoeuvre a little to get my poor memoir kept out of his hands.[19]

This was an early paper of Huxley's, before he had established his reputation. Boole found the refereeing process even chancier. He sent a paper to the Royal Society which was almost returned immediately because nobody knew him. Ultimately, it was sent out to two referees, one of whom accepted and the other rejected it. After further discussion, it was finally agreed to publish the paper. The outcome was that the published paper was later judged to be a significant contribution to mathematics, and Boole was awarded the Society's Gold Medal for it. If being unknown was a disadvantage, so being known to the editors was often an advantage. For editors did not always consider it necessary to use referees for papers from established authors. A paper that Rayleigh submitted for the British Association meeting in 1886 provides a striking illustration of this reliance on reputation. Somewhere along the line his name became detached from the document. The paper was initially rejected by the selection committee on the grounds that its results seemed inherently unlikely. They then discovered who the author was, and immediately accepted his paper. Rayleigh was well aware of the bias against new authors. In looking through the Royal Society's archives, he came across a paper from the relatively unknown John Waterston, which had been submitted in 1845, but not published. He realised that the work it contained anticipated important ideas that had appeared since, and insisted that it should be published – almost fifty years after it had first been submitted.

Perhaps of wider significance than bias in selection was the lack of organisation in the publication system. Communication between authors and editors could be poor, and who did what was often not clearly defined. Frankland encountered this problem with an important contribution to chemistry that he reported to the Royal Society:

This paper was at once ordered by the Council to be printed in the Transactions; but the Secretary, Prof. Stokes, unfortunately locked the paper in his drawer and forgot all about it for upwards of twelve months. No communication had been made to me that the paper would be published, and I had long ago concluded that it had not been thought worthy of publication, and had, consequently, been sent to the archives. This delay in the publication of my paper gave me, subsequently, a good deal of trouble, because . . . several other chemists had begun to work in the field . . . and now began to dispute my priority.[20]

Despite this lapse, Stokes was widely regarded as a reliable and hard-working editor. The latter characteristic was particularly essential. Crookes was by way of being a professional editor, and the continual stress of dealing with journals affected his health. Joseph Hooker, though a journal editor himself, was scathing about the effort involved. 'I do not see how a really scientific man can find time to conduct a periodical scientifically – or brain to go over the mass of trash that is communicated to it and requires expurgation'.[21] The plus side was the influence that an editor could exert. Thus Crookes used his editorship of the *Quarterly Journal of Science* to propagate his interest in spiritualism. Lockyer, as Editor of *Nature*, became so influential that it was suggested he was confusing himself with the Author of Nature (at the time, a common description of God).

Publishing research

One frequent problem was deciphering the manuscripts that were submitted for publication. For much of the century these were hand-written, and the text had to be read successively by the editor, the referee (if one was employed) and the printer. This left plenty of room for errors to creep in. Huxley complained bitterly: 'Your printers are abominable. They make me say that "Tyndall did not see the *drift* of my statement", when I wrote *draft* as plainly as possible'.[22] (Do not believe stories that say Victorians all had good hand-writing. Huxley's left a good deal to be desired, and he was far from unique.) Hardly surprisingly, editors welcomed the introduction of the typewriter towards the end of the century. Crookes bought one in the 1880s, when he was in his mid-fifties, and rapidly became proficient.

Another part of the difficulties was the way printers operated. Until the latter half of the century, most printers had only a limited amount of type available. This meant that an author had to return proofs as quickly as possible, before the printer needed the type for another customer. For a longer publication, such as a report or a book, the printer might set up the initial section, the author correct the proofs, and then the printer might reuse the original type to set up the next section. Things subsequently became easier as methods were developed for making moulds of each page before the type was redistributed. Printers also had the unfortunate habit of keeping the author's manuscript, so authors either had to retain a second hand-written copy, or try to remember what they had written. One of Darwin's sons described the typically careful way in which his father approached the matter:

> It was characteristic of him that he felt unable to write with sufficient want of care if he used his best paper, and thus it was that he wrote on the backs of old proofs or manuscript. The rough copy was then reconsidered, and a fair copy was made. For this purpose he had foolscap paper ruled at wide intervals, the lines being needed to prevent him writing so closely that correction became difficult. The fair copy was then corrected, and was recopied before being sent to the printers. The copying was done by Mr. E. Norman, who began this work many years ago when village schoolmaster at Down. My father became so used to Mr. Norman's hand-writing, that he could not correct manuscript, even when clearly written out by one of his children, until it had been recopied by Mr. Norman. The MS., on returning from Mr. Norman, was once more corrected, and then sent off to the printers.[23]

At the beginning of the century, printing was essentially a skilled manual process, as it always had been. Typesetting continued this way for most of the century, but other activities, such as the actual printing on paper, became increasingly automated. In general, these developments had most impact where a large number of copies had to be produced rapidly, as with newspapers. They were less important for the shorter runs that typified journal publishing. The changes of most importance for scientific journals related to the reproduction of pictures. Earlier, the choice was mainly between crude woodcuts, or detailed, but expensive, engravings on copper. In the nineteenth century, wood engraving and lithography also became available. The problem was that these methods were both costly and required the services of skilled artists. Some idea of the cost can be gained from an estimate for printing the *Transactions of the Geological Society* in the 1820s:[24]

Printing	£94/2/0
Paper	£57/8/0
Lithography	£67/18/6
Copper plates	£117/12/0
Colouring	£31/10/0
Total	£369/0/6

The last three items relate to the illustrations that were included. (The final one is for hand-colouring of the plates, after they had been printed.) The illustrations took up less space than the text, but accounted for some 60 per cent of the cost. The burden was particularly heavy for natural history journals, where good illustrations were often vital. Most societies included the cost of the journal in members' subscriptions. The Zoological Society initially followed this practice, but then found that it had to charge a guinea a year for its *Proceedings* in order to cover the cost of the illustrations. Along with cost, was the time taken by the artists to complete their task. When the Professor of Anatomy at University College London produced a new textbook in the 1860s, it took four years for all the coloured lithographs to be prepared satisfactorily.

Later in the century, the need grew to reproduce photographs in scientific journals and books. Cameras were being used for scientific recording, but the resultant photographs still had to be redone by artists in order to be printable. Direct printing of photographs would both speed up the publishing process and ensure that the details were accurately reproduced. Scientists were soon involved in investigating the problem. In 1862, De la Rue – whose family firm was interested in any new method of printing – reproduced photographs of sunspots in the Royal Astronomical Society's journal for the first time. The year before, one of Huxley's friends had illustrated a paper in the *Natural History Review* with a photograph (though Huxley, himself, continued to use traditional drawings). Subsequent improvements both in photography and in the reproduction of photographs meant that photographic illustrations became increasingly common in journals by the end of the century. Photographs, apart from their ability to record details that might be missed on an engraving, had the great virtue that the scientist was in charge of the process. Engravers generally knew little about the science they were illustrating, and their work had to be carefully monitored. (Though the Assistant Secretary of the Royal Astronomical Society, William Wesley, was also one of the best scientific engravers in London.)

Prior to photography, illustrations were usually based on the original drawings supplied by the scientist. Naturalists, in particular, needed to be competent artists. For example, Huxley's on-the-spot sketches of marine life, made on his voyage in HMS *Rattlesnake*, were later considered good enough to reproduce in the report of the expedition. Less artistically accomplished scientists sometimes had recourse to optical gadgets – especially the camera lucida. This instrument projected an image of the scene or object being viewed onto a sheet of paper, so allowing it to be traced. Airy, for example, used one to produce a drawing of Lord Rosse's giant telescope in Ireland. One problem with illustrations was that, to reduce the cost involved, they might be copied – sometimes with permission, and sometimes without – from one journal or book to another. Piazzi Smyth, a leading astronomer, compared a drawing of a scene in Tenerife, that had been reproduced more than once, with a photograph of the same scene that he had taken himself:

It is instructive as connected with the language of drawing, to trace the gradual growth of error and conventionality as man copies from man. Errors are always copied, and magnified as they go; seldom are excellences reproduced. After a few removes, the alleged portrait of nature is only a caricature of the idiosyncrasies of the first artist.[25]

Tracking down research

As the century progressed, scientific journals and their contents were increasingly organised in a standard way, to make it easier to find information. One example is the incorporation in journal papers of references to previous research in the field. In the early years of the century, this was done in a fairly haphazard way, and it was not uncommon for papers to contain no explicit citation of other work. By the end of the century, the provision of references was widespread, to the extent that one of the older generation of scientists complained: 'The quoting and reference now so extensively practised seems to be the bane of all literature'.[26] But the main complaint was rather that references had still not become standard enough:

I am referred by an author to a paper by Schmidt, in the *Berichte* of the German Chemical Society, vol. xx. Not possessing this journal, I hope to be able to find an abstract of the paper in question in the *Journal* of the Chemical Society, to which I subscribe; but as I have no notion in what year vol. xx of this *Berichte* was published I have to search through numerous indexes in order to find the abstract.[27]

The mention of an 'abstract' here is significant. As the amount of literature grew, journals tried to help their readers through the maze by printing summaries of papers appearing in other journals, more especially those published abroad. (The alternative of publishing the same paper in a number of different journals had drawbacks, not least in clogging up the literature even more.) But printing such summaries in existing journals meant that there was less space for new papers, since lack of finance limited the size of many nineteenth-century science journals. The solution was to publish the summaries separately in an 'abstract journal'. Just as journals had only printed summaries of papers relevant to their own interests, so the abstract journals were subject-based. The limitation on doing this was again money. For example, the Physical Society could not afford the cost of producing an abstract journal in physics, so it combined with the much wealthier Institution of Electrical Engineers to publish a joint abstract journal in the two fields. Victorian scientists tended to see their struggle to control the literature as a kind of military battle, which, by the end of the century, they believed they were winning. So, in the 1890s, the President of the Chemical Society commented: 'We, in this society, can never be too grateful to Professor Williamson for having led the storming party to victory which established our system of abstracts'.[28]

The problem was not simply finding out that a paper existed, but also of tracking down a copy of the journal in which it appeared. It was hoped that abstracts would help solve this by providing sufficient information, so that it became unnecessary for a reader to consult the original version. This worked to some extent, especially in chemistry, but many readers still needed to look at the original paper in order to get more details. Libraries with a good coverage of the scientific literature were essential. All the societies provided library facilities for their members; but all faced financial problems in maintaining them. Part of this was due to the expansion in the amount of relevant literature that had to be purchased as the century progressed. But the resultant need for more storage space and for more library assistance also created difficulties. Librarians were not well paid – perhaps only £100 a year –

and often had other jobs. For example, the librarian at the Royal Geographical Society in the latter years of the century was also a sub-editor of *Nature* and on the staff of *The Times*, while the librarian at the Institution of Electrical Engineers was simultaneously its secretary and the editor of its journal. Accommodation, too, often left a great deal to be desired. The Patent Office was set up in London in the aftermath of the Great Exhibition. It soon acquired a valuable science and technology library, but this was housed in a narrow corridor popularly known as the 'drainpipe'. Readers were expected to use their top hats as tables for taking notes. When the new building at South Kensington to house the British Museum (Natural History) was designed in the 1870s, the architect failed to include any library space at all. But at least scientists in London had access to a wide range of libraries. Elsewhere, access to scientific literature, and especially to journals, depended on what local bodies – such as the Literary & Philosophical societies – were active.

For really eminent scientists, the problem of having their research made widely available was solved by republishing all their papers in one or more volumes towards the ends of their lives. For example, Cambridge University Press agreed to publish Lord Rayleigh's collected papers at the end of the century. The only difficulty arose over the motto that Rayleigh wished to include at the beginning:

When I was bringing out my *Scientific Papers* I proposed a motto from the Psalms, 'The works of the Lord are great, sought out of all them that have pleasure therein'. The Secretary to the Press suggested with many apologies that the reader might suppose that *I* was the Lord.[29]

⤝ 5 ⤞

Speaking and writing

The art of lecturing

SOONER or later all nineteenth-century scientists had to lecture. Even Darwin who disliked the activity, had to present his ideas to other naturalists. Even worse, when he returned from his voyage in the *Beagle* he was pressurised into becoming Secretary of the Geological Society. This meant that he had to present not only his own work, but also the papers of geologists who could not be present at the meetings. Still, his wealth and his continuing ill health allowed him to avoid giving talks for much of his later life. The opposite was true of scientists at the Royal Institution. Their livelihood depended on attracting paying audiences, which depended, in turn, on their lecturing abilities. Indeed, the early success of the Institution was closely tied to its appointment of Davy to lecture on chemistry.

The sensation caused by his first course of Lectures at the Institution, and the enthusiastic admiration which they obtained, is . . . scarcely to be imagined. Men of the first rank and talent, – the literary and the scientific, the practical and the theoretical, blue-stockings, and women of fashion, the old and the young, all crowded – eagerly crowded the lecture-room. His youth, his simplicity, his natural eloquence, his chemical knowledge, his happy illustrations and well-conducted experiments, excited universal attention and unbounded applause.[1]

The eloquence may have been natural, but Davy studied the art of lecturing in some detail. Dalton was an experienced teacher; yet, when the older man came down from Manchester to lecture at the Royal Institution, Davy carefully instructed him in how to proceed. Dalton later explained:

Mr Davy advised me to labour my first lecture; he told me that people here would be inclined to form their opinion from it. Accordingly, I resolved to write my first lecture wholly, to do nothing but tell them what I would do and enlarge on the importance and utility of science. I studied and wrote for nearly two days, then calculated to a minute how long it would take me reading, endeavouring to make my discourse about fifty minutes. The evening before the lecture Davy and I went into the theatre. He made me read the whole of it, and he went into the farthest corner; then he read it and I was the audience. We criticised upon each other's method.[2]

Faraday, Davy's successor as the shining star of the Royal Institution lectures, was equally industrious in developing a good technique:

In early days he always lectured with a card before him with Slow written upon it in distinct characters. Sometimes he would overlook it and become too rapid; in this case [his assistant] had orders to place the card before him.[3]

Tyndall, Faraday's successor, developed his lecturing techniques as a schoolteacher. He was appointed to Queenwood, an experimental Quaker school in Hampshire and one of the first to have science laboratories. In later years at the Royal Institution, he found that a drink before lecturing improved his style. So Roscoe records that, when Tyndall came to Manchester, he asked for champagne beforehand to cater for the well-being of his brain during the lecture. It obviously worked. Armstrong subsequently attended lectures by both Huxley and Tyndall and found the latter much more inspiring. At Queenwood, Tyndall met Frankland, who was a fellow-teacher. Initially, Frankland was not a good lecturer, but practice in lecturing at Queenwood allowed him to develop an acceptable style. Not that teaching and lecturing necessarily required the same skills. Dewar was not a good university teacher, but his lecturing talents blossomed at the Royal Institution, where he was appointed professor in the 1870s. Galton felt there were too many boring lectures, and worked out a way of judging how gripping a lecture was: he measured the level of fidgeting among members of the audience.

Lecturing in universities

University scientists could hardly complain of lack of opportunity to lecture. Often they gave a range of courses with only limited assistance. For example, Maxwell, when he was at Aberdeen, was in charge of the whole of the natural philosophy course. He lectured for 15 hours a week (along with weekly oral and written examinations for all his students). Lodge, as professor in Liverpool in the 1880s, had to teach physics to all three undergraduate years and mathematics to the first two years with the assistance of one sole demonstrator. Not everybody was as conscientious in their teaching as these two. Lodge had

Faraday lecturing at the Royal Institution. Prince Albert and his sons are seated listening to Faraday on the left.
From *Illustrated London News* (16 February 1856).
The British Library, PP7611

The physicist John Tyndall giving one of his many popular lectures on science.
From *Vanity Fair* (6 April 1872).
Newspaper Library, Colindale

attended mathematics lectures at University College London a decade earlier:

Clifford was a brilliant mathematician, but he had his own methods, and he was not a systematic teacher. He was often half an hour late: it was worth waiting for, however, when he did come. Occasionally he gave us an examination; but he used to say that an examination was more trouble than a lecture, because you had to think beforehand what to ask, whereas in a lecture you could say what came into your head.[4]

Mathematicians, in particular, seemed to have problems in sticking to the laid-down lecture times. Ball recalled one of his lecturers in Dublin who was so interested in his subject that his lectures lagged further and further behind the syllabus. As the end of term approached his lectures therefore got longer and longer, until the final one of the term extended from lunch to dinner. Some of Stokes' lectures at Cambridge also ran far over their allotted time. He, however, had some justification. He was lecturing on optics, and the Sun was the only source of light for his lecture demonstrations. After spells of bad weather, he would therefore catch up by extending the length of his lecture. At the University of Cork, several staff and students caught the train to attend lectures. Unfortunately, the train timetable ensured that they were always late. Boole, one of the staff affected, suggested that the college clock should be run permanently a quarter of an hour slow to get round this problem. (He was voted down.)

Mathematicians also had their peculiarities in terms of lecturing style. J. J. Thomson attended Cayley's lectures at Cambridge. It was a small group, and they sat round a table:

Cayley did not use a blackboard, but sat at the end of a long narrow table and wrote with a quill pen on sheets of large foolscap paper. As the seats next the Professor were occupied by my seniors, I only saw the writing upside down. This, as may be imagined, made note-taking somewhat difficult.[5]

Ball's mathematics lecturer at Dublin did somewhat better for his students, though he, too, sat at the head of a table, 'having a pile of writing material, together with sheets of carbon paper, by means of which he could write several copies of his notes at once. . . . We took such notes as we could, and scrambled for the carbon copies'.[6] Some of these problems were solved by an increasing reliance on blackboards. Though the blackboard first appeared in the first half of the century, its full flourishing came in the second half. Apart from allowing more people to see what the lecturer had written, it was also an excellent way of constructing pictures to illustrate specific points. Huxley was a master of this approach: 'The course was amply illustrated by excellent coloured diagrams, which, I believe, he had made; but still more valuable were the chalk sketches he would draw on the blackboard with admirable facility, while he was talking, his rapid, dexterous strokes quickly building up an organism in our minds, simultaneously through ear and eye'.[7] So, when Huxley gave his famous lecture 'On a piece of chalk' to a British Association audience at the end of the 1860s, 'chalk' already implied 'blackboard' for his audience.

Most biomedical lecturers used large, usually coloured, diagrams hung on the walls to illustrate their talks, as Huxley did. But nineteenth-century scientists were keen users of new visual aids as they appeared. Frankland and Tyndall attended a lecture in Paris where a powerful light was used to project the view through a microscope onto a screen. Frankland observed with awe that the audience was shown cheese mites that looked as big as rats. The two introduced the idea to England on their return. Many leading scientists also illustrated their lectures with practical demonstrations. Stokes was regarded with particular admiration, for he used very simple apparatus and his experiments always worked. Such demonstrations were easier to devise in the physical sciences than in the biological; but even Lister in his surgical lectures tried to introduce one or two practical illustrations. His students particularly remembered him giving his lecture on the circulation of the blood with one hand held above his head. He eventually brought it down and showed them how different it was in colour from his other hand.

Opportunities for teaching, and being taught, science improved at all levels in the latter half of the century. By the mid-century, the Universities of London and Durham had appeared, as had Owens College in Manchester. This still left England with fewer universities per head of population than any other major country in Europe. The logjam burst in the 1870s, with several university colleges springing up in that and subsequent decades. Most appeared in industrial cities (though the Welsh colleges were more rural). The new institutions were labelled 'colleges', rather than 'universities', because they initially depended on London University for their degree examinations. In fact, it was some time before teaching students for degrees became their most important activity: the initial emphasis was more on part-time students following sub-degree courses.

These new bodies were strongly biased towards science. Mason's College in Birmingham (now Birmingham University) was originally set up specifically to teach scientific and technical subjects, and the proposed college in Bristol only received financial support from Oxford University after it agreed to teach subjects in the humanities as well as the sciences. All of the new colleges eventually covered a full range of disciplines, since this was necessary before they could become independent universities. But their initial bias towards helping local industry meant that new opportunities occurred for interested scientists. The problem was that the great amount of teaching required and the emphasis on practical applications left little time for pure research. Even so, a chemist like William Ramsay at Bristol, and a physicist like Oliver Lodge at Liverpool, were able to establish their reputations as researchers by working in the new colleges.

South Kensington

The London University examinations provided an umbrella for these new institutions, but the most important new institution of all started as an independent creation. The complex history of the institutions that grew up in South Kensington calls to mind Winston Churchill's definition of Russia: 'a riddle wrapped in a mystery inside an enigma'. So far as science was concerned, the initial development started in central London. De la Beche, who had become Director of the new Geological Survey of Great Britain, opened a Museum of Economic Geology in 1841, and recruited excellent staff – such as Forbes, Playfair and Joseph Hooker – to staff it. Ten years later, the Museum was enlarged, moved to a new site, and given the grander title of the 'Government School of Mines and Science Applied to the Arts'. (Arts in the nineteenth century meant subjects requiring practical skills.) It was now a near neighbour of the Royal College of Chemistry, set up in 1845. The College had obtained the 'Royal' title from Prince Albert, who played a major role in the events that were subsequently to move both of these institutions to South Kensington.

At the end of the 1840s, Prince Albert became involved in the idea of a 'Great Exhibition'. This would be held in London, and would show off and compare industrial innovations from countries round the world. The Prince recruited Henry Cole and Lyon Playfair to help organise the event. (Cole was a leading member of the Society for the Encouragement of Arts, Manufactures and Commerce, a body founded back in the mid-eighteenth century.) The Great Exhibition of 1851, housed in its enormously popular 'Crystal Palace', made a hefty profit.

A well-known Victorian writer Henry Mayhew and cartoonist George Cruikshank join forces to celebrate the 1851 Exhibition in London.
The British Library, Add. MS 35255, f.135

This money, plus some more from the Government, was set aside to be used by the Commissioners of the 1851 Exhibition in ways that would advance science and the arts. The first step they took was to purchase a large area of land in South Kensington for future development (it was sometimes called 'Albertopolis').

Further activity soon followed on the Government side. A Department of Science & Art was created in 1853, with Cole and Playfair as its two senior civil servants: they were called Art Secretary and Science Secretary, respectively. The department was established on the South Kensington site, since its aims were similar to those of the 1851 Commissioners. Originally, it came under the Board of Trade, emphasising the importance attached to practical applications. Subsequently, as its role in teaching increased, responsibility was transferred to the educational side of Government. Both the School of Mines and the College of Chemistry were put under the control of the Department, and, in the early 1870s, both were moved to be near it on the South Kensington site. By the end of that decade, the area was covered with a remarkable variety of institutions: some still there – like the Albert Hall – and some long gone – like the Museum of Fish-Culture, which was run by William Buckland's son, Frank. (His main concern was introducing new species of fish to Britain – and, of course, what they tasted like.)

The science side of teaching at South Kensington was initially in the hands of Huxley for biology, Frankland for chemistry, and Guthrie (who replaced Tyndall) for physics. But the person in charge, John Donnelly, was actually an officer in the Royal Engineers. The pressure to expand science education in the latter half of the nineteenth century raised the inevitable question – who will teach the teachers? In the early years of the Science & Art Department, part of the answer was to recruit from the Army. So long as there were no major crises abroad, the Army was prepared to second officers in the Royal Engineers, with their good technical grounding and organisational skills, to South Kensington. Donnelly actually stayed there for the rest of his service career, ending his days as a major-general. He was obviously the sort of person that W. S. Gilbert had in mind when he wrote *The Pirates of Penzance*:

> I'm very good at integral and differential calculus,
> I know the scientific names of beings animalculous,
> In short, in matters vegetable, animal and mineral,
> I am the very model of a modern Major-General.

That obviously confident biologist, Thomas Huxley. From *Vanity Fair* (28 January 1871).
Newspaper Library, Colindale

The problem in teaching terms at South Kensington was that the School of Science and the School of Mines, though initially linked together, had differing views of their new status. Murchison, who took over the School of Mines from De la Beche in 1855, was never happy with the transfer from Trade to Education. He believed that the Oxford and Cambridge graduates who looked after education might retard the growth of science: 'Such persons may be eminent in scholarship . . . and yet ignorant of the fact that the continued prosperity of their country absolutely depends upon the diffusion of scientific knowledge among its masses'.[8] Huxley, on the contrary, saw the move as an opportunity for a major expansion of science teaching. In the event, Huxley proved to be right, for the Science & Art Department was allowed to go its independent way until the end of the century. The two Schools were finally separated at the beginning of the 1880s, when Huxley became Dean. As H. G. Wells recorded, this title had its advantages:

My mother did not like to cast a shadow on my happiness [on becoming a student at South Kensington], but yet she could not conceal from me that she had heard that this Professor Huxley was a notoriously irreligious man. But when I explained that he was Dean of the Normal School, her fears abated, for she had no idea that there could be such a thing as a lay Dean.[9]

As Wells indicates, Huxley relabelled the science section at South Kensington the 'Normal School of Science'. 'Normal' here was used to mean that the School was primarily aimed at producing teachers, like the Ecole Normale in Paris, but the name proved unpopular. Within a few years, the School was renamed the Royal College of Science. Huxley also oversaw an expansion of teaching staff (the new additions included Lockyer). The number of students correspondingly increased, and, by the 1890s, the success of the College meant that space was at a premium. The Professor of Geology complained bitterly:

In spite of the greatest care in ventilation, the effects of the insanitary and overcrowded condition of these laboratories are beginning to make themselves painfully felt. Of the four female students two have broken down, and some of the less healthy men are beginning to be affected.[10]

As provision for science improved, attention increasingly focussed on the parallel need for better technical teaching. The wealthy City of London Livery Companies came under pressure to provide financial support for such teaching, since it was highly relevant to their professed aims. One major result was the foundation in 1881 of the first technical college in England, at Finsbury in central London. It was followed three years later by the Central Institution of the City & Guilds, which was built on the South Kensington site, and provided technical education at a higher level. The new professors included Armstrong, who lectured on applied chemistry. He figured largely in efforts to promote the City & Guilds initiative, as the description of a large reception held in 1887 indicates:

Between musical items many interesting exhibits were inspected and the Professors of the college gave lectures and demonstrations. One of the rooms of the college was especially illuminated by the Electric Power Storage Company, another by the Auer von Welsback Incandescent Gas Light. In the chemical department Armstrong lectured on *The Production of Madder Colours and Indigo from Coal*, with many interesting exhibits.[11]

The expansion of schools in the latter decades of the nineteenth century created a great demand for teachers of all subjects. Scientists who cared about the way their discipline was presented to beginners now had an opportunity, as they taught future teachers, to explore their own educational ideas. For example, some felt that, since science consisted of finding things out for oneself, teaching should follow the same pattern. Armstrong, in particular, pressed this approach. Indeed, he claimed that the usual method – where the lecturer simply presented results to students – made learning more difficult, citing Huxley's lectures as an example:

As a young student, in 1866–7, I sat, if not exactly at his [Huxley's] feet, in face of his blackboard and watched him embroider it most exquisitely with chalks of varied hue: the while he talked like a book: with absolute precision, in chosen words, so easily that we were hypnotised by his basilisk artistry into the absurd belief that we were learning: in fact, we were just being told, allowed to have no doubts, with no time to think! He was a marvellous exponent – therefore, a bad teacher, as are all who are eloquent.[12]

So, when Armstrong started teaching at the college in Finsbury in the 1880s, he decided to train his students in the way that he thought chemistry ought to be taught:

I determined to attempt to lead students to be discoverers from the outset – to train them to use their eyes and to think for themselves, on all possible occasions. . . . I simply developed the age-old method of the infant and the detective, in place of didactic teaching. I taught very little chemistry but a great deal of chemical method.[13]

The idea of 'heuristic' teaching – as this approach came to be called – was not new. Darwin's mentor, Henslow, left Cambridge to take charge of a nearby parish towards the end of the 1830s. There he developed a botany course for the local school which required the children to collect and examine plants for themselves. If anything, Henslow was more successful than Armstrong. Whereas traditional chemistry teaching, via demonstrations, only required the lecturer to have access to apparatus and chemicals, heuristic teaching meant that the pupils must also have such access. Many schools could not afford this. Collecting plants cost nothing. The general view of science teaching by the end of the century was that it should combine factual learning with some exploration: the amount depending on the subject and on local circumstances.

Examinations

To try and improve both the number and the skills of school-teachers rapidly, the Science & Art Department decided to provide a financial incentive via a new examination system. Examinations became highly popular in the mid-nineteenth century (in part, as a way of controlling nepotism and emphasising advancement by merit). Oxford and Cambridge universities, for example, both set up boards for providing schools with external and nationally recognised examinations. (These were originally referred to as examinations for middle-class schools: the title was changed to 'local examinations' when some schools protested that they were too wealthy to be considered middle class.) All these new examinations included science and mathematics, but, hardly surprisingly, scientific and technical subjects were particularly strongly represented in the examinations which the Science & Art Department started in 1859. What the Department did to encourage teachers was to introduce a system of payment that depended on examination results. For each pupil who passed one of the Department's examinations, the teacher received a payment (the exact amount depending on how well the pupil had passed). A bright student, who did well across a wide range of subjects, could provide the teacher with a useful income. The young H. G. Wells, for example, took a wide range of the examinations on offer, mainly studying the subjects via textbooks. He generally got first-class passes, each of which earned his teacher £4. What Wells got out of it, ultimately, was a training place at South Kensington. In contrast with Armstrong, Wells regarded his year in Huxley's class as the most educational of his life, and Huxley, himself, as an inspiration for his science writing.

Staff at South Kensington were involved in the Science & Art examinations both directly and indirectly, by persuading colleagues to help. So Huxley wrote to Hooker in 1864:

Donnelly told me today that you had been applied to by the Science and Tarts Department to examine for them in botany, and that you had declined. Will you reconsider the matter? I have always taken a very great interest in the science examinations, looking upon them, as I do, as the most important engine for forcing science into ordinary education.[14]

Along with the scientists, the Royal Engineers at South Kensington were pulled into the system, and often examined in technical subjects. They were needed, for the examinations became more and more popular. In 1860, just under 1,500 candidates took the examinations. By the early 1870s, this number had risen to 30,000 or more, and, by the latter part of the 1880s, it had reached 100,000. Needless to say, the Treasury became increasingly anxious about the amount of state funding that was required. Indeed, it became a significant factor in the ultimate demise of the Science and Art Department at the end of the century. But Huxley and his colleagues remained convinced that the approach worked. Huxley wrote to Donnelly in 1889, as the pressure for controlling expenditure grew:

I am unable to see my way (and I suppose you are) to any better method of State encouragement of science teaching than payment by results. The great and manifest evil of that system, however, is the steady pressure which it exerts in the development of every description of sham teaching. And the only check upon this kind of swindling the public seems to me to lie in the hands of the Examiners.[15]

External lecturing

Huxley and his contemporaries were especially keen to interest working-class men in science. Wells, whose father was a professional cricketer, fell into this category (though, through his work, Wells senior was acquainted with influential men who were also keen cricketers – Darwin's friend, Lubbock, for example). Many of those involved with the Science & Art Department also participated in giving public lectures to working-class audiences. Owen gave a series of popular lectures at South Kensington in 1858 in a lecture theatre which held 450 people. Of these, 350 were reserved for working men, their wives and children above 15 years of age. In the 1860s, attendance at a course of ten lectures by such luminaries as Huxley and Tyndall cost 5*s*. Up in Manchester, Roscoe organised a better deal: 13 lectures at 2*s* 6*d*, again including Huxley and Tyndall. Efforts to reach this kind of audience actually stretched back to the beginning of the century – to when George Birkbeck put on an annual series of free lectures in Glasgow. He later explained:

Whilst discharging the duties of Professor of Natural Philosophy and Chemistry, in Anderson's Institution, in Glasgow, I had frequent opportunities of observing the intelligent curiosity of the 'unwashed artificers', to whose mechanical skill I was often obliged to have recourse . . . the question was forced upon me, why are these minds left without the means of obtaining that knowledge which they so ardently desire, and why are the avenues to science barred against them, because they are poor?[16]

Birkbeck's initiative was successful: his lectures regularly attracted audiences of three hundred. It also triggered off similar efforts elsewhere, leading to the formation of Mechanics' Institutes up and down the country. There was soon a parallel publishing campaign, with the popular *Mechanics' Magazine* appearing in the 1820s along with cheap science books sponsored by the Society for the Diffusion of Useful Knowledge. (The latter was the creation of the politician, Henry Brougham, who was a friend of Birkbeck.) In the early decades of the century, the Mechanics' Institutes played a valuable part in disseminating science to a working-class audience. Boole, for example, depended greatly on the Institute founded in Lincoln for his self-education. But, as the century progressed, the original science orientation of the Institutes declined, while the membership became increasingly middle class. Other initiatives followed. For example, the Christian Socialist group set up a Working Men's College in London in the 1850s, of which Lubbock subsequently became Principal. In the 1860s and 1870s, universities became increasingly involved, not only via the Science & Art Department, but also via the University Extension movement. As part of this latter, universities – especially Oxford, Cambridge and London – sent lecturers to provide courses in towns and cities round the country. Several of the courses were devoted to science: indeed, Lubbock makes another appearance as chairman of the Society for the Extension of University Teaching. The problem always was finding lecturers who could hold the attention of an audience at the end of a hard day's work. Eminence was no guarantee of this. Lodge commented unhappily on one of Kelvin's supposedly popular lectures:

The pioneer anthropologist John Lubbock sitting in the House of Commons. His contemporaries thought his activities as a Member of Parliament were as important as his scientific work. From *Vanity Fair* (23 February 1878).
Newspaper Library, Colindale

'the idea may easily grow that anything by Sir William Thomson is mainly unintelligible'.[17]

Many local societies catered for an interest in science. Major centres often had a Literary & Philosophical Society, which usually included lectures on science, and might also organise field trips to study natural history, geology, or archaeology. During the nineteenth century, many local clubs specifically devoted to these sort of topics appeared. The members of such societies joined for a variety of reasons. One contemporary suggested that archaeology societies had four types of members: 'the Archaeologist proper; the Harkaeologist, who comes to listen; the Larkaeologist, who comes for the fun of things; the Sharkaeologist, who comes for the luncheon'.[18] The same was true of societies devoted to other subjects. For the most part, eminent scientists of the day were not greatly involved in the workings of these local societies, but they considered such activities well worthwhile and supported them when asked. The Revd. Charles Kingsley was an enthusiastic natural historian. When he became a canon of Chester Cathedral in the early 1870s, he started a natural history society in that city, and called on his friends in the scientific world for support. So he wrote to Lyell:

I have just started here a Natural Science Society – the dream of years. And I believe it will 'march'. But I want a few great scientific names as honorary members. That will give my plebs, who are of all ranks and creeds of course, self-respect; the feeling that they are initiated into the great freemasonry of science, and that such men as you acknowledge them as their pupils. I have put into the hands of my geological class, numbering about sixty, your new *Students' Elements* . . . These good fellows, knowing your name, and using your book, would have a fresh incentive to work if they but felt you were conscious of their existence.[19]

For people on low incomes, gaining access to science books was not always easy. The alternative to buying was borrowing. Yet most libraries with good science collections demanded the payment of a subscription. The libraries in Mechanics Institutes gave some help, and free public libraries began to appear after the middle of the century. By 1880, around a hundred such libraries were available, the great majority of them in England, and mostly in towns of some size. They provided useful access to science books across the country; though science readers in London continued to have the great advantage of free access to the libraries of the British Museum and the Patent Office. In addition, the growing demand for instruction in science led to the publication of an increasing number of cheap science titles.

Science publishing

Some publishers were especially linked with the production of science books for general consumption. Murray, who published both Lyell and Darwin, is an obvious example. Longman is another, though Tyndall thought the firm underestimated the attraction of science. (They printed only 500 copies of his *Fragments of Science*; yet the American edition of 2,000 copies sold out on the day of publication.) But the big new name was Macmillan. The Macmillan brothers, Daniel and Alexander, came from Scotland to become booksellers in England. Their shop in Cambridge was frequented by a number of scientists, including Clerk Maxwell, Galton and Geikie. After Daniel's death in the mid-1850s, Alexander set up a branch in London, where he soon encountered Huxley and his friends. Booksellers in the nineteenth century often also acted as publishers. Alexander moved down this pathway – one result being the foundation in 1869 of the science journal, *Nature*. Its editor, Lockyer, became Macmillan's adviser on science books, and the firm explored the idea of producing science texts that were easy to read, yet written by experts in the field. In the 1870s, Macmillan launched a *Science Primer* series, with the initial titles written by such authorities as Huxley, Roscoe, Balfour Stewart and Geikie. The same sort of initiative was taken by other publishers. An *International Scientific Series* was set up by an American publisher, who cooperated in Britain first with the publisher, Henry King, then, after his death, with Kegan Paul. Almost inevitably, the people turned to for advice on this international project included members of the X-Club, and they and their friends contributed titles to the series. Huxley, during this period, was regularly contributing science articles to popular reviews and magazines. A letter he wrote to the editor of one of these, the *Fortnightly Review*, might equally apply to his writing of scientific texts:

I am always very glad to have anything of mine in the *Fortnightly*, as it is sure to be in good company; but I am becoming as spoiled as a maiden with many wooers. However, as far as the *Fortnightly* which is my old love, and the *Contemporary* which is my new, are concerned, I hope to remain as constant as a persistent bigamist can be said to be.[20]

The astronomer Norman Lockyer in his heyday as editor of the journal *Nature*. In later years, he allowed his beard to grow more luxuriantly. From A. J. Meadows, *Science and controversy* (Macmillan, 1972).
The British Library, X620/2910

Huxley was far from being the only eminent scientist to write for a range of magazines. His scientific foe, Owen, also spread himself widely – from the upmarket *Edinburgh Review* and *Quarterly Review* to the popular *Hood's Magazine* and *Household Words*. He even wrote in the *Comic Miscellany* (debunking the idea of ghosts). Scientific controversies could, and did, spill over into these more popular outlets. In the early 1860s, Tait and Kelvin disputed Tyndall's ideas on heat, first in *Good Words*, and then in the *North British Review*. Ten years later, Tait and Tyndall were at it again, first in *Nature*, then in the *Contemporary Review*. Yet, throughout the century, the editors of these magazines were often dubious whether scientists could satisfy the needs of their readers. The editor of the *Quarterly Review* complained at the end of the 1830s: 'Our Whewells, Brewsters, Lyells, etc, are all heavy, clumsy performers; all mere professors, hot about little detached controversies'.[21] The editor of *Good Words* was even more damning. He commented that the Sunday reading matter in some Christian families was so dry that they might even prefer to read the science articles in his journal.

⇥ 6 ⇤

Scientists in society

Salaries

A VICTORIAN who wanted to become wealthy would not have considered a career in science. Very few scientific jobs were available in the first half of the nineteenth century. Even those that existed could have their drawbacks. Airy, for example, when he first became Professor of Astronomy at Cambridge received the title and the observatory, but very little else. As the wits of the time said, happily misquoting Shakespeare, the University gave 'to Airy nothing, a local habitation and a name'. When Faraday became Davy's assistant at the Royal Institution, he was paid 25 shillings a week together with accommodation in the building. Subsequent promotion to superintendent of the laboratories earned him an additional 5 shillings a week. Even promotion to professor only brought in £100 a year. Davy, despite marrying a wealthy wife, was sometimes suspected of cutting corners to improve his income. Babbage claimed disapprovingly that, on one occasion, 'Davy contrived to transfer between three and four hundred pounds from the funds of the Royal Society into his own pocket'.[1] The standard salary at the Royal Institution continued to be £100 into the second half of the century. This was what Tyndall was paid. It compared poorly with the £200 he had been paid as a teacher at Queenwood College.

Many dedicated scientists looked to the academic world for their support. The new universities in the latter half of the century provided jobs, but the salaries were not necessarily generous. When Frankland became professor at Owens College in Manchester, he was given £150 a year plus two-thirds of the student fees. The *Manchester Guardian* welcomed the appointment, but deplored the salary. 'While we congratulate the community upon the choice thus made, we must give expression to a feeling very prevalent, and at which we have already hinted – the inadequacy of the remuneration offered to such a man'.[2] Though such comments were obviously valid, the money from students' fees helped Frankland to survive. Frankland was on much the same level as Boole at Cork, who received altogether £300 a year from his university. Since he was supporting a wife and four daughters, even £300 left little leeway. He itemised his annual expenditure as follows: food, wages and washing – £132; rent – £36; local taxes – £12; income tax – £21; coal – £11; gas – £3; insurance – £32; garden – £2; stamps and stationery – £3; clothes – £25; self – £20; charity – £5; wine and beer – £6.[3] Things improved later in the century. When Lodge went to Liverpool in the 1880s he received £400 plus two-thirds of the student fees. After a couple of years, money from the latter source more than doubled his income. Like several of his academic contemporaries, Lodge also made money from external lecturing and examining. The figure of £400 seems to have been about the going rate for a professor then. It was what Ramsay received at Bristol as professor of chemistry. When he subsequently became Principal of the college, his salary only increased to £650 per year.

The older universities were more generous. Lister applied for the professorship of surgery at Glasgow. The salary was only £400 a year, but the main financial attraction was the opportunities it offered for private practice. Scottish universities tended to be generous. When Owen was being urged to succeed Forbes at Edinburgh, the salary mentioned was £1,000 a year. Ball received £500 a year at Dublin. This was an acceptable middle-class salary, but, as he remarked on becoming Astronomer Royal of Ireland, traditional posts often assumed more wealth than this:

Those who built the observatory appear to have thought it must be necessary for the Astronomer Royal to keep horses and carriages; at any rate, a range of stables was provided which was somewhat out of proportion to the establishment which a man of science is generally able or willing to maintain.[4]

Airy, as Astronomer Royal at Greenwich, was equally unwilling to live too grandly. In his case, it was less a lack of money than the caution with which he expended it. His position was one of the few financially supported by the state, and he ensured that it was also the best paid. He kept in close contact with the Admiralty, who were responsible for the Royal Observatory, and, by the 1870s, had persuaded them to raise his pay to £1,200 a year. Such a sum could support a family very comfortably. Lister, for example, as an eminent medical man commanded an equally good salary in the 1870s. His establishment then included a butler, a cook and a coachman, together with a number

The Astronomer Royal, George Airy, looking rather more benevolent than he appeared to some of his colleagues at the Royal Observatory. From *Vanity Fair* (13 November 1875). Newspaper Library, Colindale

of more junior staff. When Darwin and his family moved to Kent, they had some half-dozen servants, though his entertaining was relatively limited.

The problem of comparative salaries was remarked on by Lyell, when his friend, Sedgwick was asked to leave Cambridge to take charge of a parish. Sedgwick was actually earning a good salary for an academic, but it did not compare with what could be earned by a successful man in a traditional profession like the Church:

Sedgwick has declined a living of £1,100 a year offered by Brougham, on the ground of not being able to retain his professorship with it. I hardly rejoice at it, except for the cause of science. As a votary of geology, Sedgwick of course must not exchange a fellowship of £430 a year, and a chair of £200, for a cure of souls of £1,1000.[5]

Reasonable salaries were sometimes available in industry. Playfair was offered £400, rising to £600, to take charge of the chemical processes at a print works. But the German chemical industry gained such a high reputation in the nineteenth century that British manufacturers often preferred appointing German chemists, rather than British ones. Consultancy work was another possibility, but the fees were not necessarily high. One problem early on was that the definition of an 'expert' was based on the qualifying procedures of the existing professions. So it took some time before new-style experts, such as chemists, received legal recognition. Even in the 1860s, Crookes was bemoaning the difficulty of doing consultancy work in London

. . . as for respectable work as consulting (in the wide sense you employ the term) or analytical chemist, I get next to none. I have made possibly £100 in six years at that work. . . . When a great prize is to be competed for there are one hundred applicants from my class of chemists. One gets it, and is held up as an instance of the advantage of cultivating science . . . but what becomes of the 99 unsuccessful ones? They starve in places of £150 a year, or are kept by their friends, or wait in hope of getting something better next time.[6]

Eminent scientists were, of course, better placed than others for obtaining extra income as advisers. In 1881, Huxley was appointed an Inspector of Fisheries at a salary of £700 a year. He received this sum in addition to his salary at South Kensington. One of the most useful bodies in terms of consultancy was Trinity House, which was mainly concerned with maintaining navigational aids round the coasts. Since these, especially lighthouses, underwent considerable development during the century, outside advice was regularly sought. Faraday was an adviser earlier in the century. He was followed by Tyndall, who was followed by Rayleigh. Tyndall initially received £100 a year from Trinity House. By the time he resigned in the 1880s, this had risen to £400. One official post had a particular cachet for scientists – that of Master of the Mint. Making coinage might seem some distance from science, but it actually involves a good knowledge of metallurgy. For a nineteenth-century scientist, however, the chief attraction was the prestige of being appointed to a post that Isaac Newton had held. John Herschel was made Master of the Mint in 1850. He was followed by Thomas Graham, who had previously been professor of chemistry at University College London. Graham's period of office was not entirely happy, and, on his death, the Mint was reorganised, removing the need for the person in charge to have a scientific background.

Other sources of money

Of course, financial problems loomed largest for those scientists with no family backing. Marrying a wealthy wife – as Playfair did – or inheriting and investing wealth – as Darwin did – made life much easier. Most eminent scientists had become reasonably well off by the time they died, but those who started wealthier usually ended that way. Galton, for example, left over £60,000 in his Will. For the inventive, patentable ideas could sometimes produce valuable income. So Crookes' radiometer – a spin-off from his research – brought him in useful royalties when it was marketed as a toy. (It consisted of a glass bulb containing vanes which spun round when the bulb was exposed to sunlight.) A number of scientists and engineers who made money from their inventions retired early and concentrated henceforth on research. This was true, for example, of W. H. Perkin, who retired on the money he derived from manufacturing the first synthetic dye, and of James Nasmyth, who devoted himself to astronomy after retiring on the proceeds from inventing the steam hammer (he left a quarter of a million pounds in his Will).

One of the most popular ways of earning extra money was by publishing for a general audience – both books and articles in magazines. To gain an adequate income from magazines required the writing of many articles. Scientists often could not find the time for this. Crookes noted that he only obtained about £20 a year for the articles he wrote, whereas he obtained three-and-a-half guineas a week for editing a popular journal. A book that made a major impact could be a much more profitable activity. When Davy produced his book on *Elements of Agricultural Chemistry* in the early years of the nineteenth century, he received 1,000 guineas for the first edition together with 500 guineas for each succeeding edition. In the 1830s, Buckland received £2,000 for each edition of his Bridgewater Treatise on *Geology & Mineralogy*. Todhunter published a stream of mathematics textbooks from 1853 onwards which were so successful that he was able to resign his college fellowship at Cambridge. Hooker produced a *Primer of Botany* in 1876 which remained in print till 1909. He remarked that it was the most profitable of all the activities he had undertaken.

A rather chancier way of gaining income was to angle for a state pension (not to be confused with a retirement pension). Dr. Johnson's dictionary says of such a pension: 'In England it is generally understood to mean pay given to a state hireling for treason to his country'. Robert Peel had some success in reforming the system when he attained political power in the early decades of the nineteenth century. The basic purpose – to help men of learning and science to carry on their work – could obviously benefit scientists who had fallen on hard times. For example, when Wallace's investments in railways and mines failed, Darwin and Huxley persuaded Gladstone to give him a pension of £200 a year. But pensions might be spread more widely. For instance Boole's wife was given an annual pension of £100 after his death to help her survive. That, at least, had some justification, but pensions were also awarded to people who did not really need the money. So, when Airy agreed to become Astronomer Royal, an annual pension of £300 was shortly afterwards bestowed on his wife to help with the family finances.

Government – advice and honours

Pensions were far from the only area where politicians and scientists made contact. Another was on various government investigations. For example, Playfair was recruited to a Royal Commission inquiring into sanitary conditions in towns in the 1840s. Frankland was similarly recruited to a Commission set up towards the end of the 1860s, which looked at water supply and pollution. Both appointments made good sense, since their chemical expertise was relevant to the investigations. Because the number of scientific experts was small, however, scientists might find themselves involved in less obvious activities. Airy, as one of the few government scientists, pops up all over the place. He was, for example, a major influence on the Railway Gauge Commission in the 1840s, which recommended the demise of the broad gauge tracks introduced by the Great Western Railway along with standardisation on the narrower gauge in use elsewhere in the UK. Although Faraday was not a government scientist, he was also seen as an all-purpose help. A letter he wrote to the Admiralty, when he was suffering from ill-health, illustrates the demands made on him:

I have had the honour and pleasure of applications, and that very recently, from the Admiralty, the Ordnance, the Home Office, the Woods and Forests, and other departments, all of which I have replied to, and will reply to as long as strength is left to me.[7]

Such investigations could sometimes be hazardous. Faraday and Lyell were asked to investigate a mining accident in the 1840s. As part of their work, they had to go down the mine to see what conditions were like there, and what safety precautions were being taken.

Faraday asked in what way they measured the rate at which the current of air flowed in the mine. An inspector took a small pinch of gunpowder out of a box, as he might have taken a pinch of snuff, and allowed it to fall gradually through the flame of a candle which he held in the other hand. His companion, with a watch, marked the time the smoke took going a certain distance. Faraday admitted that this plan was sufficiently accurate for their purpose; but, observing the somewhat careless manner in which they handled the powder, he asked where they kept it. They said they kept it in a bag, the neck of which was tied up tight. 'But where', said he, 'do you keep the bag?' 'You are sitting on it', was the reply.[8]

The biggest scientific jamboree of them all was played out in the 1870s. British exhibits had done well in the Great Exhibition of 1851, but industrial competitors overseas improved rapidly. By the end of the 1860s, there was considerable concern that British industry was now lagging behind. Various commentators suggested that a major cause was the neglect of scientific training and research in Britain. The pressure built up, and, in 1870, the government set up a Royal Commission on Scientific Instruction and the Advancement of Science. William Cavendish, the seventh Duke of Devonshire was appointed president. The list of members includes other familiar names – Huxley, Lubbock, Stokes – with Lockyer as secretary. The Commission dealt with every conceivable aspect of the subject – from science in schools to government support of research – over the course of its five-year existence. Unfortunately, its numerous recommendations were mainly ignored. One reason was that the Gladstone administration, which had set up the Commission, fell before it could deliver its final report.

By this time, scientists were being regularly honoured for their work, as the steadily increasing number of scientific knights indicates. Many of the names that have been mentioned in previous chapters ultimately bore the prefix 'Sir'. John Herschel tried at first to decline, on the grounds that such an elevation might interfere with his scientific work. Murchison, on the contrary, exerted much energy in seeking a knighthood. In the 1840s, when his campaign was at its peak, he hosted a series of great soirées to emphasise his central role in the scientific scene. When Peel, at first, turned down his request for a knighthood, Murchison changed his political affiliation. Lyell observed: 'Murchison is one who has worked at science chiefly for the rewards'.[9] By the end of the century, science was even infiltrating coats of arms. Crookes included four prisms and the inevitable radiometer in his. Some scientists climbed even further up the ladder. Kelvin was knighted in 1866 for his work on the Atlantic telegraph, and made a peer, for his scientific work as a whole, in 1892. Such elevation cost money. Owen received a knighthood and might have received a higher honour, but he found the fee required was more than his annual salary. Airy typically refused a knighthood twice on the grounds that it cost too much. He eventually agreed to accept one when his salary was increased. Some names are conspicuous by their absence – Darwin, Huxley and Tyndall, for example – were never knighted. However, evolutionary zeal was not necessarily the sole obstacle here. After all, Hooker, as keen an evolutionist as anyone, was knighted for his work.

The award of such honours was always slightly arbitrary, and sometimes even confused. Brewster described his own experience in the 1830s under King William IV. Initially, Brewster turned down a knighthood on the grounds that he could not afford the fee of £109. But he came down to London when he was told that the fee would be waived. He handed over his card, which bore the words, 'To be knighted'.

On presenting it, however, the lord in waiting exclaimed that he knew nothing of it – a previous intimation to him having been forgotten. An awkward moment ensued, but my father said quietly, 'Let us move on'; which was answered by the King's exclamation, 'No! no! I know, I know!' Having no sword, he borrowed the Duke of Devonshire's, and with that performed the usual ceremony.[10]

The physicist Lord Kelvin: clearly a man who has done well out of his scientific ventures. From *Vanity Fair* (29 April 1897). Newspaper Library, Colindale

As the papers in his hand suggest, Lyon Playfair was better known to his contemporaries for his political activities on behalf of science, rather than for his work as a chemist. From *Vanity Fair* (20 February 1875). Newspaper Library, Colindale

Brewster was not the only one to face problems. Playfair was thought to have been knighted in 1885, but, owing to confusion at the ceremony, this was subsequently found to be untrue. It made little difference, for, in 1892, he was raised to the peerage. Unlike Kelvin's elevation, Playfair's was tied up with contemporary politics. Playfair left the Science & Art Department towards the end of the 1850s to become professor of chemistry in his old department in Edinburgh. Though he still spent much time in London advising the government, he clearly missed being at the centre of things. The Universities of Oxford and Cambridge (and Trinity College Dublin) had long elected their own Members of Parliament. At the end of the 1860s, this right was extended to the Scottish universities and to the University of London. Playfair immediately stood as a candidate for the one of these new seats, and, on being elected, gave up his chair and moved back to London.

Politics

British political parties underwent considerable change in the nineteenth century; although the main division was, as in the previous century, between the right-wing Tories and the more liberal Whigs. During the course of the century, these two groups gradually transmuted into the Conservatives and the Liberals, but there were various hiccups on the way. One was provided by the conservative Sir Robert Peel and his followers in the first half of the century. They regarded themselves as liberal conservatives, and this put them at odds with the old-style Tories in the House of Commons. Another was the split in the Liberal party over home rule for Ireland in the second half of the century. The opponents of home rule, led by Chamberlain, became the Liberal Unionists, and they found themselves increasingly siding with the conservatives.

Playfair was a life-long Liberal, and he contributed in various ways to Gladstone's successive administrations (though the position he seems to have enjoyed most was that of Postmaster-General). When the Scottish university voters turned against his politics, he was elected by the working-class constituency of South Leeds. At the 1885 election, his fellow Liberal candidate for North Leeds was another scientist, Arthur Rucker. (Rucker lost by a narrow margin, and the following year moved to South Kensington to be professor of physics.) At the same election, Roscoe was returned as the Liberal Member of Parliament for South Manchester. Lubbock was already at Westminster. He initially failed to be elected for his West Kent constituency – it was said due to his Darwinian leanings. Shortly before the 1865 election, he published one of his major works, *Prehistoric Times*. Darwin wrote to say he had read the book with interest, but he foresaw the outcome. His letter ended: 'I do sincerely wish you all success in your election and in politics; but after this last chapter, you must let me say: "Oh dear! Oh dear! Oh dear!"'.[11] Lubbock finally became the Liberal M.P. for Maidstone in 1870.

The fact that all these stood in the Liberal interest was no coincidence. There were eminent scientists on the Conservative side – for example, Stokes became a Conservative M.P. for Cambridge University – but the balance among scientists was definitely on the Liberal side. Darwin's family was traditionally Whig back into the eighteenth century. Charles was upbraided by his sisters for daring to send them copies of the Tory magazine *John Bull* from Edinburgh, even though they knew he was joking. Towards the end of the 1870s, Gladstone visited Darwin, with Playfair, Lubbock and Huxley in tow. 'Darwin commented afterwards, with genuine pleasure: "What an honour that such a great man should come to visit me"'.[12] Hooker was a moderate Whig, like Darwin, but both Huxley and Tyndall were more radical. In the mid-1880s, Gladstone disrupted this comparatively simple picture by committing the Liberal party to home rule for Ireland. (The parliaments of Great Britain and Ireland had been united since 1800.) Some scientists – Playfair and Roscoe, for example – remained in the Liberal party, but others – including Tyndall, Huxley, Lubbock and Kelvin – joined the splinter National-Liberal group. Both Tyndall and Kelvin (who became one of the leaders of the National-Liberal group in Scotland) had been born in Ireland. So their concern is understandable. But many non-Irishmen also saw the home rule proposals as a betrayal of trust. Huxley certainly did, and it served to increase his already strong dislike of Gladstone. 'It is to me a grave thing', commented Huxley, 'that the destinies of this country should at present be seriously influenced by a man, who, whatever he may be in the affairs of which I am no judge – is nothing but a copious shuffler, in those which I do understand'.[13] Joint opposition to Irish home rule brought the National-Liberals and the Conservatives together. It led to the rather improbable result that Tyndall was approached by the Conservatives to see if he was prepared to stand as a candidate for parliament. (He was not.)

The scientists who did become Members of Parliament concentrated, hardly surprisingly, on educational and technical matters. Lubbock and

THE SCIENTIFIC VOLUNTEER.

"If ever I have to choose .... I shall, without hesitation, shoulder my rifle with the Orangeman."—*See Professor Tyndall's Reply to Sir W. V. Harcourt.* "*Times,*" Feb. 13, 1890.

A cartoon of the physicist John Tyndall, who joined the National Liberals in opposing home rule for Ireland. Loyalists had orange as their distinctive colour, which explains why Tyndall has an orange impaled on his helmet. From *Punch* (22 February 1890).
The British Library, PP5270

Playfair, in particular, were concerned with legislation relating to the medical profession. But the scientific net spread widely. Consider, for example, Roscoe and parliamentary drains:

Great complaints had for many years arisen that foul smells were noticed in the various rooms both of the Lords and the Commons. A committee, of which I was the chairman, was appointed to go into the whole question. It is scarcely credible that on inquiry it was found that there were not any plans of the drainage of the Houses of Parliament to be found . . . the fact being that when Sir Charles Barry had finished the building there was some kind of quarrel between him and the Office of Works, and the authorities were never able to get from him a plan of the drains. The consequence was that when we came to examine into the drainage, we had to cut up the thick bed of cement on which the Palace was built . . . The condition of things was frightful: in many places there was no fall; there were cess-pits under the House of Commons from which the air for the Chamber was obtained. These pits were filled with foul matter, and, in short, the state of things inside the building was about as bad as it could be.[14]

For the most part, nineteenth-century Prime Ministers were tolerant of science rather than interested in it, though they were often acquainted with scientists at the individual level. Murchison and Lord Palmerston, for example, were old friends. The two who took more than a casual interest were, perhaps surprisingly, both Conservatives. (This assessment excludes Gladstone, who certainly took an intellectual interest in science, but mainly to counter some of the claims of its adherents.) Peel knew little about science, but saw the value of its applications. On one occasion he invited his tenants and neighbours in to hear Playfair and Buckland discuss science and agriculture. Peel's interest had been stimulated by Liebig's work in Germany on agricultural chemistry, which he had heard of from Playfair. Liebig figured in the portrait gallery that Peel kept of people he considered intellectually significant. It also included portraits of Buckland and Owen. (It was Peel who, somewhat controversially, appointed Buckland Dean of Westminster.) Lyell sat next to Peel at a dinner in 1839, and afterwards commented:

He is without a tincture of science, and interested in it only so far as knowing its importance in the arts and as a subject with which a large body of persons of talent are occupied. He told me he was one of the early members of the British Association.[15]

The other Conservative Prime Minister with an interest in science was Lord Salisbury, whose governments alternated with Gladstone's over

DRESSING FOR AN OXFORD BAL MASQUÉ.

"THE QUESTION IS, IS MAN AN APE OR AN ANGEL? (*A Laugh.*) NOW, I AM ON THE SIDE OF THE
ANGELS. (*Cheers.*)"—Mr. DISRAELI's *Oxford Speech, Friday, November* 25.

Disraeli, one of the less science-oriented prime ministers, contributes to the debate on Darwin's
evolutionary ideas. From *Punch* (10 December 1864).
The British Library, PP5270

the last twenty years of the nineteenth century. Rayleigh, who was related to Salisbury by marriage, was a little dubious. He visited Lord Salisbury in 1870, and subsequently remarked:

He took me into his laboratory which also serves as a dressing room and showed me some magnetic experiments which I am supposed to explain! He is too awkward to succeed well as an experimenter I think.[16]

But Salisbury took his experimenting seriously. He engaged a Fellow of the Royal Society, Herbert McLeod , to help him with his work, and he was, himself, elected a Fellow of the Society at the end of the 1860s. He took particular delight in being elected President of the British Association in 1894, and devoted his Presidential Address to evolution. Salisbury accepted the idea of evolution – indeed, he had recommended Darwin for a knighthood in 1866, when evolutionary controversies were at their height – but, as a physical scientist, he saw the strength of criticisms relating to the timescale required for evolution. Kelvin, who had been pressing the same point, was greatly impressed by the address, whereas Huxley dismissed it as 'an awful hash'. At least Salisbury was generally agreed to be sympathetic to science and scientists. He was Prime Minister at the time of Queen Victoria's Jubilee in 1897, and Crookes, Frankland, Huggins and Lockyer all received knighthoods as part of the celebrations.

Like scientists in Parliament, scientists as a group normally only involved themselves in politics when a matter of immediate interest was concerned. There was, for example, a long debate during and after the Devonshire Report on the need for government funding to support scientific research. A notable exception to this rule was the Governor Eyre controversy. Eyre was Governor of Jamaica in 1865 when there was an uprising amongst the population. He suppressed the revolt with considerable brutality, sparking off an angry debate in Great Britain over whether or not his actions had been justified. Murchison, who knew Eyre, helped organise an Eyre Defence Committee, but found himself opposed by a phalanx of biologists. As Herbert Spencer reported, the anti-Eyre Jamaica Committee:

. . . was remarkable for containing all the leading evolutionists – Darwin, Huxley, Wallace and myself, besides others less known. Indeed, the evolutionists, considering their small number, contributed a far larger proportion to the committee than any other class.[17]

Lyell was also anti-Eyre, but Hooker and Tyndall were more sympathetic to the governor, who eventually was not prosecuted.

Interestingly, the pro-Eyre group included several of the leading writers of the day – Carlyle, Dickens, Kingsley, Ruskin and Tennyson.

Literary interests

This division between scientists and writers over the Eyre campaign was unusual: more often, they overlapped in their views. Indeed, both groups were interested in each other's activities. Writers frequently mentioned scientific ideas, and even sometimes the scientists themselves. W. H. Mallock's novel, *The new republic*, appeared in the 1870s. It poked fun at a range of intellectuals of the time, including the scientists Huxley and Tyndall and the mathematician, W. K. Clifford. Similarly, individual scientists – Owen, Huxley, Lockyer and others – were from time to time examined wryly in *Punch*. Dickens regarded science with amused misgivings. He was acquainted with Owen, and mentions his fame as an anatomist in *Our mutual friend*, where he describes Mrs. Podsnap as 'a fine woman for Professor Owen' because of her 'quantity of bone'. Ruskin was both more interested in, and more knowledgeable about science. He had come under Buckland's influence at Oxford, and was made Secretary of the Geology Section of the British Association when it met in Oxford in 1847. Dickens, by way of contrast, preferred to poke fun at the British Association via his reports of the Mudfog Association for the Advancement of Everything.

Carlyle was different. His dislike of material progress might be expected to have put him at odds with the scientists; but he was redeemed in their eyes by his emphasis on the importance of individual genius in influencing the course of history. Nineteenth-century scientists saw the growth of their discipline as dominated by a limited number of outstanding practitioners. Carlyle was proud of his mathematical abilities, and applied at one stage for the chair of astronomy at Edinburgh University. Darwin was not totally impressed. He reminisced about, 'a funny dinner at my brother's, where, amongst a few others, were Babbage and Lyell, both of whom liked to talk. Carlyle, however, silenced everyone by haranguing during the whole dinner on the advantages of silence'.[18] Owen was on friendly terms with Carlyle, but not with Carlyle's wife. Jane Carlyle said that his sweetness reminded her of sugar of lead (a sweet, but poisonous compound). Henceforth, Carlyle's scientific opponents nicknamed him 'sugar of lead'. But the scientist most smitten by Carlyle was, rather surprisingly, the highly materialistic Tyndall. The two became so close that, when

Carlyle's wife died, Tyndall took him for a holiday in the South of France to help him get over it.

But in terms of closeness to members of the scientific community as a whole, Charles Kingsley and Alfred Tennyson certainly head this list. Kingsley had always been keen on natural history, and, as an undergraduate at Cambridge, he had joined Sedgwick's geological field trips. Later, he wrote on natural history, and when *Nature* was launched, Lockyer invited him to review natural history books for the new journal. Kingsley was one of the few clergymen to welcome Darwin's evolutionary ideas with enthusiasm. Correspondingly, he was one of the few clergymen whom Huxley was prepared to stomach. 'He is', Huxley wrote, 'a very real manly, right minded parson but I am inclined to think on the whole that it is more my intention to convert him than his to convert me'.[19] Four years after *Origin of species* appeared, Kingsley summarised his view of it in *The water babies*. This was widely read and approved of by members of the scientific community. Kingsley was acquainted with both Huxley and Owen and provided a composite portrait of the two of them in his book, disguised as Professor Ptthmllnsprts (i.e. 'Put them all in spirits' – the usual method of preserving biological specimens).

Tennyson had been one of Whewell's favourite students at Cambridge, and he maintained a strong interest in science throughout his life. References to science abound in his poems, and he so impressed contemporary scientists with his interest in science that he was elected to the Royal Society. (When Tennyson died, Kelvin was one of the pall-bearers at his funeral.) Tennyson's preferred science was astronomy, though he was somewhat hampered in pursuing it by his bad eyesight. He was on friendly terms with both Pritchard and Lockyer. The latter records some characteristically Tennysonian remarks as they observed the heavens together:

One night when the moon's terminator swept across the broken ground around Tycho [a large lunar crater] he said 'What a splendid Hell that would make'. Again after showing him the clusters [of stars] in Hercules and Perseus he remarked musingly, 'I cannot think much of the county families after that'.[20]

For an earlier scientific generation the key poet was Wordsworth. As a young man, Davy encountered the poet, Robert Southey, who introduced him to Coleridge, who, in turn, introduced him to Wordsworth. They all had a high regard for Davy, despite his interest in chemistry. (After William Godwin had dined with Davy, he told Coleridge that it was a pity such a man was degrading his talents by studying chemistry.) Davy's poetic judgement was held in sufficiently high regard that Wordsworth and Coleridge sent him copies of the poems to be included in the second edition of *Lyrical ballads* for copy-editing. Whewell, too, was well acquainted with Wordsworth. He came from the part of the Lake District where Wordsworth lived, and Wordsworth's brother was Master of Whewell's college at Cambridge. Airy came from the same area and also knew Wordsworth. He noted after a tea in 1841 that: 'Mr Wordsworth is as full of good talk as ever'. Though Wordsworth had his hesitations about science, his emphasis on nature was widely popular among scientists. The journal, *Nature*, first appeared long after Wordsworth's death, but his writings provided its motto: 'To the solid ground / Of nature trusts the Mind that builds for aye'. The significant difference was that the journal gave the capital letter to 'Nature', while 'mind' was spelt with a small 'm'.

Darwin was a constant reader of Wordsworth: indeed, like many nineteenth-century scientists, he was an avid reader of all sorts of literature, including both poetry and novels. Perhaps he appreciated the latter particularly because so many Victorian novels dealt with matters of inheritance. Like Faraday, he enjoyed both being read to, and reading aloud himself. Not long after his return from the *Beagle* voyage, we find him wondering why reading the work of the philosopher, Comte, gave him a headache, whereas reading Dickens' *Sketches by Boz* cured it. Hooker exchanged notes with Darwin about the books they were reading. In the period immediately after the *Origin of species*, he was particularly enthusiastic about the novels of George Eliot. (There was a slight embarrassment here, for she had for a number of years been enamoured of their mutual friend Spencer.) Faraday was another keen reader, and he communicated his interest to those around him. His niece remembered that the highly religious Faraday nevertheless took great pleasure in reading Byron's poetry.

In the 1860s, Rayleigh kept a record of what he had been reading as an undergraduate. Apart from work books, it was mainly novels, including Wilkie Collins' *The woman in white*, Charlotte Bronte's *Shirley* and *Jane Eyre*, and Trollope's *Framley parsonage*. The Wilkie Collins and the Trollope were quite recent publications, but the two Bronte titles had been around for over a decade. Whewell commented on the two of them to his wife when they had not long been published:

Shirley is, I think, much cleverer and more dramatic than *Jane Eyre*. Then it puzzles me much as to the sex of the writer. It has even more of the cleverness, largeness of speculation and audacity, which made me think *Jane Eyre* a masculine performance; but then there are some ways of dealing with male and female relations which look like feminine workmanship. For instance, all the women fall in love with the men, which is I think a female characteristic. But if it be a woman's book, women are growing to be very strange and alarming creatures.[21]

Brewster actually met Charlotte Bronte. He was in London for the Great Exhibition of 1851 and recorded:

One of the most interesting acquaintances I have made since I came here, I made yesterday. It was that of Miss Bronte, the authoress of *Jane Eyre* and *Shirley*, a little, pleasing-looking woman of about forty, modest and agreeable. I went through the Exhibition with her.[22]

Brewster was on friendly terms with Walter Scott, as too was Davy, for his wife was one of Scott's cousins. In later years, Tait went one better: he had Robert Louis Stevenson as a student. This created some problems:

Stevenson's father was Thomas Stevenson, the well-known lighthouse engineer. He hoped that his son would carry on the family traditions, and expressly desired [that he] work with optical apparatus. But the future essayist and writer of romances had not the smallest elementary knowledge of the laws of reflexion and refraction. The immediate purposes of the Physical Laboratory were lost on him; though no doubt what little training he allowed himself to undergo bore some fruit when a few years later he read a paper before the Royal Society of Edinburgh comparing rainfall and temperatures of the air within and without a wood. It was published in the *Proceedings*: literary critics have, however, left it severely alone.[23]

Scientists as literary men

Scientists were, of course, themselves writers. Indeed, Huxley' popular writings were frequently held up as an example of good, readable prose. But their concern was with non-fiction. Babbage did consider writing fiction when he was short of money: 'I proposed to give up a twelvemonth to writing the novel. But I determined not to commence it unless I saw pretty well clearly that I could make about £5000 by the sacrifice of my time'.[24] He dropped the idea when he found how much initial capital was required for publication. When scientists did indulge in literary expression, it was usually in poetry. In 1839, John Herschel decided it was time to dismantle his father's famous large telescope, as it was deteriorating badly. When the scheduled day arrived, he assembled his family inside the telescope tube and read them a poem he had written as a farewell to the telescope. Nor was poetry writing confined to the first half of the century. Huxley wrote a poem on the train as he returned from Tennyson's funeral in 1892, and it was subsequently published in the journal *Nineteenth Century.*

Southey, Coleridge and Wordsworth all believed that Davy would have made an excellent poet. Davy, like most scientists who tried their hands at serious verse, was particularly influenced by Wordsworth. However, he realised better than most that the Wordsworthian approach could sometimes prove too pedestrian. He mapped out a parody that began:

> As I was walking up the street
> In pleasant Burny town
> In the high road I chanced to meet
> My Cousin Matthew Brown.[25]

William Rowan Hamilton in Ireland, though of a younger generation than Davy, also fell under Wordsworth's spell. In his case, Wordsworth was less impressed. He wrote to Hamilton: 'You send me showers of verses which I receive with much pleasure, as do we all: yet have we fears that this employment may seduce you from the path of science'.[26] Whewell wrote a sonnet about Wordsworth for the latter's daughter. The initial lines run:

> Daughter of that good man whose genuine strain
> Patiently uttered oft in evil days
> Called English poesy from erring ways
> Of laboured trifling, insincere and vain[27]

Wordsworth was not the only model: Sylvester preferred Dante. He described the progress on one of his sonnets to Lockyer:

... now it has received the very highest polish of which it is susceptible. It is a fraction reduced to its lowest terms – i.e. it has been reduced to the most simple form of expression capable of conveying the meaning – it is now simple and direct as if it had come from the hands of Dante who I think would not have been ashamed to claim it as his own.[28]

Sylvester was trying to persuade Lockyer to publish the sonnet in *Nature*, but, by the 1880s, when Sylvester wrote this, serious poetry was no longer seen as entirely appropriate for scientific publications.

Humorous scientific verse and parodies remained in favour throughout the century. In the 1840s, while at Cambridge, Galton set up a society for writing short, humorous poems. One writer of such verse was, almost inevitably, Buckland at Oxford. In the early years of the century, it was still believed that toads embedded in stones could survive without food for long periods. Buckland put the belief to the test, and summarised his results in verse:

> Toad did you not a promise give
> Without one meal one year to live
> O wicked toad then tell me why
> In spite of this you dared to die.[29]

Airy was opposed to giving public money to Charles Babbage to assist with the construction of the latter's computer (or 'calculating engine' as it was then called). He scoffed at Babbage's progress in a parody of the children's poem, 'This is the house that Jack built', substituting the refrain, 'This was the engine that Charles built'. But it was Maxwell who was the acknowledged master of scientific humour in poetry. His parody of Burns' poem, 'Gin a body meet a body', still appears in anthologies:

> Gin a body meet a body
> Flyin' through the air,
> Gin a body hit a body,
> Will it fly? And where?[30]

Such versifying of scientific ideas – in this case, on how bodies collide – was typical of Maxwell. He and Tait even exchanged ideas on how to carry out experiments in verse. For example, Maxwell advised Tait:

> Take, then, a coil of copper pure
> And fix it on your whirling table,
> Place the electrodes firm and sure
> As near the axis as you're able
> And strive to learn the way to work it
> With galvanometer in circuit.[31]

But Maxwell also enjoyed writing about his scientific peers. For instance, Lockyer and his work on the Sun's atmosphere:

> And Lockyer, and Lockyer,
> Gets cockier, and cockier
> For he thinks he's the owner
> Of the solar corona.[32]

James Clerk Maxwell, the physicist, in the years preceding his early death. From W. D. Niven, *The scientific papers of James Clerk Maxwell* (Cambridge University Press, 1890). The British Library, 8706.h.21

He equally enjoyed parodying their ideas. Thus one poem was inspired by Tyndall's presidential address to the British Association in 1874, when Tyndall caused considerable controversy by stressing his belief in materialism. Maxwell commented with amusement (in verse that sounds like a preview of Kipling):

In the very beginnings of science, the parsons, who managed things then,
Being handy with hammer and chisel, made gods in the likeness of men;
Till Commerce arose, and at length some men of exceptional power
Supplanted both demons and gods by the atoms, which last to this hour.[33]

It is often forgotten that the Revd. Charles Dodgson (Lewis Carroll) was a fringe member of the nineteenth-century scientific community: his mathematical interests actually overlapped with those of Boole. Nor was he alone in writing both children's stories and light verse; so did his fellow-mathematician, William Clifford. Carroll's parodies and humorous verse were widely appreciated within the scientific community, but they were, unfortunately, rarely devoted to science. One slight exception was *Hiawatha's photographing*, where he described the chemical processes involved in photography:

> Finally, he fixed each picture
> With a saturate solution
> Of a certain salt of Soda –
> Chemists call it Hyposulphite.
> (Very difficult the name is
> For a metre like the present,
> But periphrasis has done it.)[34]

Hobbies

Many of the scientific community were, like Dodgson, interested in photography. Indeed, the 'hyposulphite' that he mentions came into use as a consequence of John Herschel's early investigations into photography. Previous attempts at photography had failed to come up with a good way of making photographic images permanent. Herschel not only showed how this could be done, but also produced one of the first properly fixed photographs. This was of his father's telescope, taken a few months before the family gathered within the tube to bid it farewell. Herschel similarly reflected the interest of many in the scientific community in his love of music. His father had been one of the leading musicians in England, and Herschel, himself, played the violin and the flute. When in London, he frequently went to concerts. One of the people he met was Felix Mendelssohn, whose piano playing he greatly admired. Owen was an even keener attender at musical events. Weber's opera, *Oberon*, was performed 31 times at Covent Garden in 1826, and Owen went to all 31 performances. Owen was a good singer and played both the cello and the flute, but many scientists who were less accomplished musically enjoyed listening to music – Darwin and Hooker, for example. When Darwin was feeling depressed towards the end of his life, he was cheered by a visit from the pianist, Hans Richter, who played to him. Babbage, ever enterprising, created and rehearsed a spectacular ballet of his own. The outcome is best described by his own summary: 'The philosopher [i.e. Babbage] writes a ballet – Its rehearsal – Its high moral tone – Its rejection on the ground of the probable combustion of the opera-house'.[35]

Out of doors, some continued the leisure activities of their youth – fishing, for example. Pritchard, hard at work, wrote to his friend, Huggins: 'I do envy you and your wife fishing and sketching by the banks of the Thames. If I don't get some fishing, or something like it soon, I shall be doubled up'.[36] Another gentle pursuit that many took up was golf. Lockyer was a keen golfer and wrote a small book on *The rules of golf*. Though the rules in force in St. Andrews were agreed to be pre-eminent, there were many local variations, and Lockyer's book helped the drive towards standardisation. Tait in Scotland was even more devoted to the game. His son, Freddie, was one of the leading amateur golfers in the latter part of the century. A story about the two of them has become a part of golf mythology. It is said that Tait senior calculated how far a golf ball could be struck; his son then went onto the golf course and hit a ball considerably further. In fact, Tait senior calculated how far a ball could be struck if it was not spinning. He then pointed out that spin should take the ball a good deal further, as his son demonstrated. Following this line of thought, Tait subsequently designed a club with a specially grooved head to impart maximum spin to the ball. The German scientist, Helmholtz, visited St. Andrews and saw golf for the first time. Tait insisted that they had a round together. Helmholtz subsequently wrote to his wife:

Tait knows of nothing else here but golfing. I had to go out with him; my first strokes came off – after that I hit either the ground or the air. Tait is a peculiar sort of savage; lives here, as he says, only for his muscles, and it was not until Sunday, when he dared not play, and did not go to church either, that he could be brought to talk of rational matters.[37]

A rather more strenuous exercise – at least, as practised in their younger days – was walking. Hooker would happily walk up to sixty miles a day. His fellow-botanist, Bentham, was equally keen: 'walked many a time fifty miles a day without fatigue, and kept up five miles an hour for three or four hours'.[38] Such hardiness was not limited to botanists. The mathematician, Cayley, reminisced: 'When under twenty, have walked twenty miles before breakfast; when about thirty-two, walked forty-five miles; dined and danced till two in the morning without fatigue'.[39] Both Hooker and Cayley became interested in the even more strenuous activity of mountaineering. Initially, the attraction of the Alps was that they offered an opportunity to study glaciers, whose properties were a matter of controversy in the mid-nineteenth century. Tyndall's first book, *The glaciers of the Alps*, was published in 1860. Subsequently, the attraction for a number of scientists – including several members of the X Club – was purely the pleasure of mountaineering. The sport was not without its dangers. Francis Balfour, who Huxley thought was destined to succeed him as a leader of the biological world in Britain, died in a mountaineering accident at the age of 31. Tyndall was climbing in Switzerland when: 'I was accosted by a guide, who asked me whether I knew Professor Tyndall. "He is killed, sir", said the man, "killed upon the Matterhorn." I then listened to a somewhat detailed account of my own destruction'.[40] What had happened was that Edward Whymper, who Tyndall knew, had just led a party to the top of the Matterhorn for the first time. Unfortunately, on the way down, four of the party had fallen to their deaths.

Clubbing

Mountaineering started out as an informal activity, but rapidly led to something more formal – the creation of the Alpine Club – the Victorian love of clubs ensured that. Besides their scientific meeting places, Victorian scientists met together with each other, and with non-scientists, in a variety of ways. The wealthier members of the community often organised receptions or dinners for friends and acquaintances. Babbage was famous for this, as was Lyell. Lubbock preferred to hold breakfast parties. Even the less wealthy entertained regularly. Huxley held open house on Sunday evenings, when any of his friends were welcome to drop in for a chat and a meal. Lockyer did the same, but on Wednesday evenings. In his case, these meetings were smoking parties. He kept a rack of clay pipes for regular attenders, with their names on them. One was reserved for Tennyson, who was a dedicated smoker. Tyndall reported that when he visited Tennyson he found, 'an apparatus on the chimney-piece like a test-tube stand, in which 15 or 20 pipes were stuck. The drawers of the table were full of tobacco of various kinds'.[41] Huxley, himself, did not smoke a pipe till he reached the age of fifty, but then he became a total addict. Rayleigh, as a non-smoker, was unusual.

Of course, these invitations worked both ways; eminent scientists were often to be found at gatherings arranged by non-scientists. Tyndall, for example, recorded attending a small dinner party given by Arthur Stanley, the Dean of Westminster. The other guests included Browning, Carlyle, Froude (the historian) and Parnell (the Irish nationalist). Some held regular meetings. Alexander Macmillan, for example, hosted gatherings on Thursday evenings, mainly intended for his authors and advisers. Huxley, Spencer and Lockyer often attended, as did Kingsley, Tennyson and Holman Hunt. (The Pre-Raphaelite painter was on friendly terms with a number of his scientific contemporaries.) The group meetings came to be called the 'Tobacco Parliament', because of the debates (which were often helpful to Macmillan in formulating his publishing plans) and the number of smokers involved. Macmillan once commended Tennyson with the words: 'He smokes like a good Christian'.[42]

Some meetings were more formal. At the end of the 1860s, the debate over science and religion raged fiercely. The editor of the *Contemporary Review* suggested to Tennyson and Pritchard that some kind of meeting place was needed where leading figures in the debate could present their viewpoints. This led to the formation of the Metaphysical Society, which brought together a remarkable range of Victorian thinkers. Huxley, Tyndall, Clifford and Lubbock (who became president of the society) were there, but so were the Archbishop of York, the Dean of Westminster, Cardinal Manning, Gladstone, Ruskin, Froude, and many others. Such *ad hoc* groupings were not unusual in the Victorian age: they helped ensure that science remained a part of intellectual life in Britain throughout the nineteenth century. But the most important meeting place of minds was provided by that major Victorian institution – the club. Though London had numerous clubs, the important one for Victorian intellectuals was the Athenaeum. The Athenaeum had been founded in the 1820s – with Davy as one of the founding members – and its importance grew as the century progressed. Its rules gave preference, in terms of election, to people distinguished in art, science, or letters, and scientists who were already members took care to make sure that more were elected. In 1858, Murchison proposed Huxley for election, and subsequently told him:

I had a success as to you that I never had or heard of before. Nineteen persons voted, and of those eighteen voted for you and no one against you. You, of course, came in at the head of the poll; no other having, i.e. Cobden, more than eleven.[43]

Because of the range of membership, the Athenaeum was a good place for politics (with a small 'p', as well as a large one). In the 1880s, Welby, who was permanent secretary at the Treasury, was on bad terms with Donnelly, the man in charge at South Kensington, over what he regarded as excessive expenditure by the Science & Art Department. The scientists at South Kensington decided it was important for the future to try and smooth over the split. They therefore persuaded Welby, who was a member of the Athenaeum, to back Donnelly for membership of the club. Lockyer wrote triumphantly to Huxley to report Donnelly's election, adding:

I am more pleased than I can say as this ought to bring Donnelly and Welby closer together – and this way science lies – and I have begged Donnelly to write to Welby to thank him because Welby's proposal was I think of great importance.[44]

In both these cases, nobody voted against: the important thing in club elections was to avoid being blackballed. Scientists in the Athenaeum were quite prepared to vote against candidates they disliked, as well as to vote for those they did want as members. St. George Mivart was a well-known biologist – he was a member of the Metaphysical Society – who, however, had doubts regarding Darwinism. Huxley and Hooker combined to make sure that he was never elected to the Athenaeum.

So how did Darwin fit into this picture of social life? Darwin was elected to the Athenaeum (at the same time as Dickens) while he was still a young man. After his illness and removal to Kent, he mixed a good deal less in society – certainly much less than most of his friends – though he always welcomed guests to his home. Not all the scientific visitors were naturalists. Rayleigh, for example, stayed with him in 1870. Like Rayleigh, Darwin lived on inherited wealth (which he invested shrewdly). By the latter years of the century, this distinguished him from many of the younger scientists. A geologist, who had suffered from the poor prospects of a career in science, commented: 'Darwin is an enviable man. A pleasant place, a nice wife, a nice family, station neither too high nor too low, a good moderate fortune, and the command of his own time'.[45] Similarly, though Darwin's books about his research were surprisingly popular, he did not write them with one eye on the cash return, as many of his fellow-scientists did. Nor was he much involved in political activity, though he agreed with the majority of scientists in siding with the Liberal cause. Even so, it was unusual for a scientist of his eminence not to receive some kind of honour from the government. Almost certainly those in government who did feel he should be honoured also felt that his name would prove too controversial, if it were to be put forward. So, in terms of involvement in its social life, Darwin, in his later years, can better be described as a flying buttress of the scientific establishment, rather than a central pillar – unlike most of his friends.

⇥ 7 ⇤

Scientists abroad

Naval voyages

DARWIN'S voyage in the *Beagle* lay at the heart of his scientific career. But the obvious question is – why was it thought appropriate for a naval vessel to take a naturalist to distant parts of the world? In the aftermath of the Napoleonic wars, many British ships were laid up, but a sizeable fleet remained that had to be employed somehow. John Barrow, one of the two most senior civil servants at the Admiralty, came up with a suggestion:

To what purpose could a portion of our naval force be, at any time, but more especially in time of profound peace, more honourably or more usefully employed than in completing those details of geographical and hydrographical science of which the grand outlines have been boldly and broadly sketched by Cook, Vancouver and Flinders, and others of our countrymen?[1]

The activities that Barrow was proposing were intended not only to improve the safety of navigation of British ships, but also to help expand knowledge of sea routes around the growing British empire and to protect the commerce involved. (During the nineteenth century, the British merchant navy expanded enormously.) Barrow was elected to the Royal Society while Joseph Banks was President, and was greatly impressed by what Banks had achieved as a naturalist with Cook in the South Seas. So he envisaged that some, at least, of the planned voyages of exploration should be accompanied by scientists. One of the earliest expeditions was in 1818: it was part of a series that tried to find a North-West passage round Canada into the Pacific. The leader was John Ross, and the scientist he took on board was an army officer, Edward Sabine, who was an expert on the Earth's magnetism. This seemed more useful than a knowledge of natural history, since the north magnetic pole was known to lie somewhere round the region they were exploring. This kind of activity speeded up at the end of the 1820s, when Barrow was instrumental in having Francis Beaufort appointed to the post of

Hydrographer of the Admiralty. Beaufort was also a Fellow of the Royal Society; though he had recently left the Society's Council after a row with the President. Once he was settled in, expeditions and surveying work grew rapidly.

Which leads back to the question of what the *Beagle* was doing. The ship had already been used to chart part of the South American coastline. Robert Fitzroy was now taking it back to continue the survey. Fitzroy's special interest was in meteorology – he later became the first head of the Meteorological Office – and this voyage of the *Beagle* was to see the first official use of Beaufort's scale for measuring wind speeds. Since the ship would be near land for much of the time, it seemed an appropriate voyage for taking along a naturalist. And so came the fateful letter from Beaufort to Henslow:

Captain Fitzroy is going out to survey the southern coast of Tierra del Fuego, and afterward to visit many of the South Sea Islands, and to return by the Indian Archipelago. The vessel is fitted out expressly for scientific purposes, combined with the survey; it will furnish, therefore, a rare opportunity for a naturalist, and it would be a great misfortune if it should be lost.[2]

At the end of the 1830s, James Ross, the nephew of John Ross, was commissioned to explore Antarctica. Like Sabine, he was an expert in terrestrial magnetism, but not in natural history. He therefore took Joseph Hooker with him as assistant surgeon and naturalist. The senior surgeon was Robert McCormick, who had also held that post on the *Beagle*. McCormick wished to make a name as a naturalist (though Darwin privately thought he was an ass), and Hooker saw him as an obstacle to his own hopes of scientific advance from the voyage. He was equally disappointed by the facilities available on board ship:

Except for some drying paper for plants, I had not a single instrument or book supplied to me as a naturalist – all were given to me by my father . . . not a single glass bottle was supplied for collecting purposes, empty pickle jars were all we had, and rum as preservative from the ship's stores.[3]

Huxley's turn came a few years later. Owen Stanley, the captain of the *Rattlesnake*, was an experienced hydrographer, who was sent out to survey the area from New Guinea to the north-east coast of Australia. Huxley, though similarly designated assistant surgeon, was not plagued by on-board competition as Hooker was. He agreed with Hooker, however, about the Admiralty's lack of interest in providing facilities for anything except surveying:

Richard Owen, the anatomist, looking more like a Dickensian rogue than an eminent scientist. No doubt his opponents regarded this cartoon as a good likeness. From *Vanity Fair* (1 March 1873). Newspaper Library, Colindale

It is necessary to be provided with books of reference, which are ruinously expensive to a private individual, though a mere dew-drop in the fitting-out of a ship . . . A hundred pounds would have supplied the *Rattlesnake*; but she sailed without a volume, an application made by her captain not having been attended to.[4]

About the mid-century, the Admiralty, accepting that all the ships they controlled could add to scientific knowledge, decided to issue a manual which would indicate the kind of information to be collected. It was edited by John Herschel and the list of contributors contains a familiar roll-call of names. Herschel, himself, contributed a chapter on meteorology; Airy wrote on astronomy, Sabine on magnetism, Whewell on tides, Darwin on geology, De la Beche on mineralogy, Owen on zoology, and Hooker (Senior) on botany. The manual, the Admiralty said, was aimed primarily to give encouragement: 'to the collection of information upon scientific subjects by the officers, and more particularly by the medical officers, of Her Majesty's Navy'.[5] But the manual soon came into more general use by travellers overseas.

In the second half of the century, as scientists gained in confidence, greater attention was paid to their needs. The pre-eminent example of this is the *Challenger* expedition in the early 1870s. In the late 1860s, Charles Thomson, professor of natural history at Edinburgh, together with an influential friend at the Royal Society, William Carpenter, persuaded the Admiralty to allow them to use naval vessels for short trips to trawl the ocean floor. The results of these trawls were sufficiently interesting for Thomson, backed by the Royal Society, to ask that a naval vessel should be specifically assigned for a scientific study of the oceans round the world. After much consideration, the Admiralty agreed, and HMS *Challenger* was selected for the task. This time, the ship was converted to the requirements of the scientists. All but two of the guns were removed, and as much space as possible was converted to purpose-built laboratories and storage space. The departure of the *Challenger* in December 1872 (for some reason, all these expeditions seemed to start in the winter) is generally seen as marking the beginning of the science of oceanography. Thomson noted cautiously when he returned: 'The somewhat critical experiment of associating a party of civilians, holding to a certain extent an independent position, with the naval staff of a man-of-war, has for once been successful'.[6]

By this time, the Admiralty was being called on for assistance by a range of scientists, not least by astronomers. In the latter part of the century, many astronomers were interested in studying the atmosphere

Scientific travellers: in this case visiting India to observe a solar eclipse. From *Illustrated London News* (10 January 1872).

of the Sun. The problem was that the fainter parts of the atmosphere could only be studied at solar eclipses, when the Moon blacked out the body of the Sun. Such eclipses took place every few years, and were only visible over restricted portions of the globe – the position changed with each eclipse. Transport not only for observers, but also for their equipment, was essential, and the British astronomical community put pressure on the Admiralty to provide this. The Admiralty was often reluctant, but did help on occasion. For example, agreement was finally reached to send two naval vessels to the Mediterranean for an eclipse in 1870 (yet another December expedition). Apart from transport, the observers had to pay their own way, so the call went out for volunteers. One who responded was Tennyson, as did a number of scientists in fields other than astronomy. Lockyer led a party including both Clifford and Roscoe aboard HMS *Psyche*. The ship foundered on a rock near Naples, fortunately without loss of life or equipment. Huggins led another party, that included Crookes and Tyndall, to Algeria aboard HMS *Urgent*. They were joined by the French astronomer, Janssen, who had just escaped by balloon from a Paris besieged by the German army. Crookes kept a record of the things he took with him: they provide an illustration of what the well-equipped scientist was expected to have on an expedition. The list of medicines included chloral and chloroform (and perhaps his brandy should be added under this heading). But a lot of the space was taken up by clothing, including a best dress suit and a common dress suit, six white shirts, four night shirts and two flannel shirts. Allowance for local customs led to the inclusion of a fez and a pistol.

Other activities

Not all expeditions depended on the Navy. Wallace and Bates went to Brazil in a small trading vessel. Wallace returned to England in a similar ship after four years, but it caught fire en route, and he lost most of his collection. Bates learnt from Wallace's experience. When he returned after eleven years, he divided his collection into three parts, each sent back on board a different ship. By the time Wallace left England again on his expedition to the Far East, he had become a recognised naturalist. This allowed him to approach Murchison, who used his numerous contacts to get Wallace a much more satisfactory passage on a government ship. Wallace and Bates started collecting on the basis of advice from the British Museum, and that institution, especially after

Owen went there in the mid-1850s, constantly sought to expand its own collections. Owen particularly emphasised the need for fossil specimens. It was said that, 'there was hardly a village or township or homestead in Australia where, if anything curious happens to turn up, that the suggestion is not at once made that it be immediately sent to Professor Owen'.[7] This international trade involved considerable sums of money: at the top end, a really good dinosaur skeleton might fetch as much as £1,500. Nor was the British Museum alone. Apart from the overseas competition to buy good specimens, the number of museums in Britain increased rapidly in the nineteenth century. By the 1860s, Britain was spending less on higher education than its main competitors, but it was spending more on museums than they.

Not all collectors brought back dead specimens. The Zoological Society needed animals for display in its 'Zoo', set up at the end of the 1820s. (The word 'Zoo' rapidly came into general use as an abbreviation of its full name – the 'Zoological Gardens'.) In later years, the Society often relied on Owen for advice on what to buy, and at what price. The Zoo soon became popular with the public, as well as with members of the Society. Soon after its foundation, Edward Lear could be found there, sketching the parrots for a new publication. When Huxley felt depressed in the 1850s, Spencer took him for afternoon walks at the Zoo in order to cheer him up. Living specimens of foreign plants were equally popular. Kew Gardens and the Royal Horticultural Society, in particular, dispatched collectors to bring back live plant specimens for cultivation in Britain. For example, David Douglas went to North America in the early nineteenth century, and discovered many plants that have since become familiar in gardens. He found so many new conifers – including, of course, the Douglas fir – that he joked to Hooker (Senior) that the latter would think he was manufacturing them. Hooker (Junior) later went out collecting for Kew in North India. He noted both the problems and the material rewards of such work:

We collected seven men's loads of this superb [orchid] for the Royal Gardens at Kew, but owing to unavoidable accidents and difficulties few specimens reached England alive. A gentleman, who sent his gardener with us to be shown the locality, was more successful. He sent one man's load to England on commission and, although it arrived in a very poor state, sold for £300 ... Had all arrived alive they would have cleared £1,000. An active collector, with the facilities I possessed might easily clear from £2,000 to £3,000 in one season.[8]

The Hunterian Museum of the Royal College of Surgeons: for many years Richard Owen's favourite hunting ground. From *Illustrated London News* (4 October 1845).
Newspaper Library, Colindale

While the Navy was busy surveying the world's coasts, the Army was busy making measurements inland. Much of this work was done by the Royal Engineers, who were noted for the range of skills they commanded. Thus Captain Henry Scott was an innovator in the design of maps, but he also found time to design the Albert Hall in South Kensington. Out in the field, the survey of India pioneered by George Everest was followed up throughout the century by the Engineers. When a detachment of Royal Engineers was sent to Canada in 1858, the Secretary of State for the Colonies explained to them: 'You are going to a distant country, not, I trust, to fight against men, but to conquer nature'.[9] But the Royal Engineers did not have a monopoly of the Army's interest in science. The Royal Artillery, too, supported scientific activity, as the career of Edward Sabine demonstrates. About the same time that Sabine became involved in exploration to find a North-West passage round Canada, he began a series of measurements round the world to try and determine the shape of the globe. These used a sensitive pendulum devised by a fellow-officer, Henry Kater. Sabine, like Kater, became an influential member of the Royal Society, and this led him on to his most important work. He persuaded the Society and the military that it was important to develop a better understanding of the Earth's magnetism. Magnetic observatories were set up in various British colonies in the latter part of the 1830s, with Sabine coordinating the work. This led at the mid-century to the discovery that the Sun could have a major effect on the Earth's magnetism. Sabine was subsequently knighted, became President of the Royal Society, and ended his army career as a general.

The Army tended to rely on its own resources for manpower, partly because much of the work involved staying in the same area for an extended period. Some civilian scientists did the same. Wallace established his headquarters in Singapore in the mid-1850s, and stayed in that general area for eight years. But the need for prolonged stays occurred most often in astronomy. William Herschel had carried out a vast survey of the northern heavens. John decided that he must extend his father's survey to the southern hemisphere, so covering the whole sky between them. He therefore took his family and his equipment out to the Cape of Good Hope in 1834, remaining there for four years. The Cape was a major stopping point for ships, whether they were going on to the Indian Ocean or the South Atlantic. The *Beagle* visited: Darwin met Herschel for the first time, and was highly impressed. Indeed, the Cape was so obviously the right place for astronomy in the southern

The biologist Alfred Russel Wallace looking particularly benevolent after all his travels.
From A. R. Wallace, *The wonderful century* (Chapman and Hall, 1908).
The British Library, 08708.b.6

hemisphere that a permanent observatory had been set up there in 1820. A new director of this observatory, Thomas Maclear, arrived a few days before the Herschels. He had been trained as a doctor, and could provide them with medical advice. He was followed out by an assistant, Charles Piazzi Smyth.

Piazzi Smyth's father was a naval officer (ultimately an admiral) who had been involved in surveying activities in the Mediterranean after the Napoleonic Wars. He had become interested in astronomy, and struck up an acquaintance with the leading Italian astronomer of the time, Giuseppe Piazzi – whence came his son's name. Later on, Piazzi Smyth became Astronomer Royal for Scotland, but continued to travel extensively. Indeed, when his observatory in Edinburgh was under investigation, he pointed out how much of his own money he had spent on scientific voyages, and claimed that they should have been supported by government funds. One of his trips had been to Egypt, where he had convinced himself that the ancient Egyptians had used a standard unit of measurement when building the pyramids. His detailed measurements were duly appreciated, but his deductions were not. Such flights of fancy were not regarded as a good argument for official support for his travels. James Simpson, the Edinburgh surgeon who first used chloroform as an anaesthetic, commented sadly on, 'Professor Smyth whose measurements were most exact but whose logic was most wretched'.[10]

Continental trips

Grand Tours of the Continent – popular in the eighteenth century – were still in vogue in the early nineteenth century. They were meant to introduce young men to the joys of civilisation: those few who had scientific interests used the opportunity to learn more of science. In the 1820s, James Forbes went to Italy, where he developed an interest in volcanoes, and then to Switzerland, which stimulated his life-long interest in glaciers. Not long after, Lyell also went on tour. In his case, his interest in geology was already established, and he went out accompanied by Murchison. The tour went well, the major worry being Murchison, who was taking regular doses of drugs:

On one occasion we were on an expedition together, and as a stronger dose was necessary than he had with him, I was not a little alarmed at finding there was no pharmacy in the place, but at last went to a nunnery, where [the Mother Superior] hoped my friend would think better of it, as the quantity would kill six Frenchmen. Murchison was cured, and off the next morning, as usual.[11]

Murchison was a keen traveller, especially to Russia, but, more importantly, he was at the centre of both the geological and geographical communities not only in Britain, but throughout the British Empire. In consequence, as exploration uncovered new features, they stood a good chance of being named after him. In North America alone, there were Murchison Island, Murchison Promontory, Murchison River, Murchison Sound, Murchison Glacier, Mount Murchison and two Murchison Capes. Doubts grew during the nineteenth century whether geography was a science at all. Murchison, a founder and frequent President of the Royal Geographical Society, did not help with his flamboyant public relations activities. At one stage, he allowed a reporter from the humorous magazine *Punch* to attend the Society's meetings. Yet the Royal Society still contained a number of members interested in geography. Galton explored parts of Africa as a young man, and was much involved with the Royal Geographical Society on his return. In the 1850s, he took the lead in publishing a handbook for travellers (with suggestions from Captain Fitzroy, Admiral Smyth and others), which was read by virtually all Victorian explorers.

Tours to increase a scientist's knowledge were clearly valuable in observational sciences, such as geology and natural history. But physical scientists, too, turned to travel, in order to meet and interact with their fellows in other countries. So Davy, as soon as the Napoleonic Wars and his personal finances would allow, went off to Paris to meet the leading French scientists and exchange information. (He actually discovered a new element during his visit.) Faraday, though with him in a menial role, also managed to establish useful scientific contacts both in France and later in Italy. Davy so much enjoyed trips abroad that, towards the end of his life, he wrote a book entitled *Consolations of travel*. Most tours of France started with Paris. Herschel visited there in 1824, and wrote to his mother: 'within eight hours after my arrival in Paris, I found myself in company with almost everyone I wished to see: Arago. M. Laplace . . . Baron Humboldt . . . M. M. Thenard and Gay Lussac the chemists, Poisson and Fourier the mathematicians and many more'.[12] Whewell was there three years later. He, too, met the leading scientists, but also noted that there were other advantages of such a visit. He was, he told his aunt, 'feasting my body in a moderate

Roderick Murchison, the geologist, a man as much at ease in society and politics as in science.
From *Vanity Fair* (26 November 1870).
Newspaper Library, Colindale

degree upon French wines and French cookery'.[13] Not all scientists were equally keen on visits to the Continent. Joule believed British sanitation to be so much better than continental facilities, that he preferred to stay at home.

One virtue of a prolonged stay abroad was that it provided the opportunity to learn the language. Kelvin, for example, while still a boy, was sent to both France and Germany to improve his language skills. It did not always work. Brewster enjoyed visiting Paris, but admitted that he spoke, 'the most awful French you ever heard'.[14] For a few in the science community, languages came easily without the need for a trip abroad. Hamilton in Dublin was able to read Latin, Greek and Hebrew at the age of five; at eight he could read French and Italian; by the time he was eleven, he was studying Arabic and Sanscrit. At the ripe old age of 14, he wrote a letter in Persian to the Persian Ambassador when the latter visited Dublin. The German language is absent from this list. At the beginning of the century, when Hamilton was a boy, French science, and hence the French language, had top priority, Within a few decades, the rise of German science was apparent. The difficulties created by the Napoleonic wars led to a rethink of the German educational system. Wilhelm von Humboldt, the brother of the leading German scientist, Alexander von Humboldt, was in charge of this reorganisation, and especially stressed the need for universities to explore new knowledge. This meant, in practice, that people seeking posts in German universities needed to develop research skills.

As scientific research and publication in Germany grew, so did the need to learn the German language: by the latter part of the century, it was becoming essential for scientists in most branches of science. Rayleigh, as a young man, was told that it was essential to read German in order to keep up with research in physics. Naturalists equally felt under pressure. Darwin told Hooker that he had begun to learn German. Hooker replied: 'Ah, my dear fellow, that's nothing; I've begun it many times'.[15] Many scientists felt that the difficulty of learning German was compounded by the obscure way in which some German scientists wrote. Huxley observed: 'As men of research in positive science they are magnificently laborious and accurate. But most of them have no notion of style, and seem to compose their books with a pitchfork'.[16]

The easiest way of absorbing both the language and the research was obviously to work at a German university for a while. This became increasingly popular once German universities had adopted the idea of awarding a doctorate for properly supervised research projects. Work for a Ph.D. not only provided practical training of a type superior to anything available in Britain for much of the century; it also gave a widely recognised guarantee of ability to do research. One of the first people to follow this path was Playfair. In 1839, he went to Giessen to study under Liebig. Giessen was a small university, but Liebig and his students had given it a world-wide reputation in chemistry. Playfair arrived there just after Liebig had received new funding for the expansion of his laboratories, and soon found himself undergoing a crash course in the German language:

Liebig at this time was writing his great book on agricultural chemistry, and he invited me to translate it into English from the manuscript. I arranged with English publishers to do this for a hundred pounds, the first money which I had yet earned. My knowledge of German was not good, but I had a motive to work hard, and my translation progressed as fast as the manuscript.[17]

Frankland and Tyndall went out to Marburg University together in 1848. Lectures started at 7 a.m., followed by work in the laboratory. According to Tyndall, these activities, along with analysing results and reading the scientific literature, often took up 16 hours of the day. He commented: 'During the semester everybody appears to be gathering knowledge, it will take years of devoted effort to bring England up to the same standard'.[18] Frankland was the first Englishman to obtain a Ph.D. at Marburg – Tyndall was awarded his a little later – and he remarked on the surprise this generated locally:

[Englishmen] were generally considered to be more or less mad, which was not, perhaps, to be so much wondered at, since the only specimen the Marburgers had hitherto seen, so far as I could ascertain, was a Mr. Magnus, who was sent here to study, but was led about as a lunatic by his keeper.[19]

British students often went to work under German academics who were known to their own mentors in Britain. Frankland had studied with Kolbe, one of Liebig's disciples, in Marburg. When Armstrong went to Germany in the 1860s, he therefore joined Kolbe, who had by this time moved to Leipzig. Both Frankland and Tyndall had studied under Bunsen at Marburg. When Roscoe went to Germany in the mid-1850s, Bunsen had moved on to Heidelberg, so Roscoe joined him there. Heidelberg remained a mecca for English students for the rest of the century. Jocelyn Thorpe, who went there to do a Ph.D. in chemistry during the 1890s, found that the English colony was large enough for cricket matches to be played. Even so, British visitors noted that

German academic life had its own idiosyncrasies. Roscoe, for example, almost found himself involved in a duel. More pleasantly, when Lister visited Germany, he was greeted by medical students singing a specially written 'Carbolsaure Tingel-Tangel' (meaning more or less 'The carbolic acid rag'). Overall, the value of a period spent studying abroad was accepted by the British scientific community. Its importance was highlighted by Lodge, who was one of those who did not go:

Whether I was wise to stay seven years in London, instead of studying abroad, may be doubted. That is again among the things one finds it difficult to decide. It probably prevented my attaining first rank as a physicist.[20]

In the 1820s, Babbage, like Herschel, wended his way across the Channel. In 1828, he was in Berlin visiting Humboldt. He found that Humboldt had convened a meeting of scientists from across Germany, and was in the middle of making the necessary arrangements. Babbage was pulled in to help:

One of the first things, of course, was the important question, how were they to dine? A committee was therefore appointed to make experiment by dining successively at each of the three or four hotels competing for the honour of providing a table d'hote for the savans. Humboldt put me on that committee, remarking, that an Englishman always appreciates a good dinner.[21]

Links with the United States

Babbage's report of the Berlin meeting was an important stimulus to the subsequent creation of the British Association for the Advancement of Science. Just as Babbage had attended the German meeting, so the annual meetings of the British Association were often attended by scientists from other countries. The meeting at Oxford in 1860 became notorious for the debate between Huxley and the Bishop of Oxford. But one of the main speakers invited to participate in that debate was a visitor from the United States, John Draper. Unfortunately, Hooker found his presentation totally boring: 'A paper of a Yankee donkey called Draper on "civilisation according to the Darwinian hypothesis" or some such title was being read, and it did not mend my temper'.[22] Darwin's favourite contact in the United States was actually the leading botanist there, Asa Gray. Darwin confided to him two years before *Origin of species* was published: 'As you seem interested in the subject, and as it is an *immense* advantage to me to write and to hear *ever so briefly*, what you think, I will enclose (*copied* so as to save you trouble in reading) the briefest abstract of my notions as to *means* by which nature makes her species'.[23] Gray subsequently visited Darwin in England, and adopted one Darwinian characteristic for his return to America – a bushy white beard. Hooker later visited the United States and botanised with Gray. One of Hooker's most vivid memories, however, was of an encounter with Brigham Young, the leader of the Mormons:

Today we called on Brigham Young and had a chat with him. He is about 70, stout, well dressed, and with a refined countenance. He reminded me of a stout, elderly and thoroughly respectable butler, more than anything else.[24]

Overseas scientists were usually impressed by the welcome they received in Britain, and not least by the range of social events to which they were invited. Simon Newcomb, a leading American astronomer, visited England in 1870, and described one informal gathering:

A feature of London life that must strongly impress the scientific student from our country is the closeness of touch, socially as well as officially, between the literary and scientific classes on the one side and the governing classes on the other. Mr. Hughes invited us to make an evening call with him at the house of a cabinet minister ... where we should find a number of persons worth seeing. Among those gathered in this casual way were Mr. Gladstone, Dean Stanley, and our General Burnside ... Then and later I found that a pleasant feature of these informal 'at homes', so universal in London, is that one meets so many people he wants to see, and so few he does not want to see.[25]

Tom Hughes, famed as the author of *Tom Brown's schooldays*, was a good contact for visiting Americans. He was not only pro-American, but also well-connected in the political, literary and scientific worlds. He was a member of the group round Alexander Macmillan, and had an especially close link at the time with Lockyer. John Draper was in England again in 1870. He described attending a more formal gathering – the annual Royal Society dinner – to his son, an up-and-coming astronomer in the USA:

Among a great many lords and earls we had the Prime Minister Mr. Gladstone and the Lord Chancellor who both made speeches. Previously, I was introduced to them, and received with the utmost kindness. About 250 persons were at the table and it so happened that my lot was cast at table with a very pleasant group, Mr. De La Rue ... and Mr. Lockyer. ... He told me that he is coming to America year after next and I am truly glad that you should have the opportunity of seeing him.[26]

Lockyer paid his promised visit to the States, and enjoyed the mixture of discussion and outdoor life that greeted him there. Thomas Edison wrote to him later saying: 'I hope you will come over here again (after you have become well smoked up in London) with several other deep and mighty intellects we will take to the Mountains for a grand hunt'.[27] In the latter part of the century, as scientific life began to expand in the United States, British scientists increasingly spent time in North America. Their visits often included lectures, since these could pay well, and involved no problem with the language. When Tyndall arrived in Boston in the early 1870s to start a lecture tour, he was greeted with the news that prayers were being said in the city for his salvation. Despite (or because of) this, his lecture tour was a great success. He cleared some £2,500 in profit, though he donated it to set up three university fellowships at US universities. His tour took in, as did the tours of other writers, a discussion with his American publishers. The problem at the time was that American publishers were not bound by British copyright conditions, so US pirating of material published in Britain was always a threat. The easiest way of protecting against such theft was to have a book published in parallel in Britain and the United States. The visit that Tyndall paid was to Appleton in New York, a publisher who used the 'Young Guard' of British science as advisers and authors, in much the same way that Macmillan did in London. Huxley also spent time with Appleton when he visited the USA. If anything, Huxley was received with even greater rapture than Tyndall. It was subsequently reported that: 'his visit was so far like a royal progress, that unless he entered a city disguised under the name of Jones or Smith, he was liable not merely to be interviewed, but to be called upon to "address a few words" to the citizens'.[28]

This growth of interest in North America was reflected by the British Association's activities. In theory, the British Association was always expected to have an imperial orientation of some sort; its original objectives included: 'to promote the intercourse of those who cultivate science in different parts of the British Empire with one another'. During its first decades, such promotion was mainly confined to relevant research activities, or to hosting colonial visitors at the annual meetings. However, in 1884, the Association accepted an invitation to hold its annual meeting in Montreal. This first overseas meeting of the Association attracted the remarkable number of 750 participants from Britain. Rayleigh was the president that year, and, like many others, took the opportunity to visit the USA. He noted resignedly of his trip to Boston: 'We arrived here this morning after a night journey attended with the usual railway accident'.[29] Some of Rayleigh's contemporaries saw the USA as a potential source of jobs when none were forthcoming in Britain. The funding of science by wealthy American business men helped increase the attraction of the country. Joseph Henry, one of the most influential scientists in the USA, was often approached for advice. In the mid-1870s, he wrote to Lockyer when one such business man, Mr. McCormick, was thinking of donating money to the University of Virginia: 'Like most men of plenty of money and no scientific knowledge, Mr. McCormick has a good deal of reverence for a great scientific name, especially from the other side of the ocean'.[30] The University of Virginia had already attracted one famous name from Britain back in the 1840s, when it appointed Sylvester as professor of mathematics. Unfortunately, Sylvester became involved in an argument – apparently about the use of the English language – and thought that he had killed his opponent in the ensuing struggle. He fled the country, and remained away, even though it transpired that he had only inflicted a superficial wound. However, thirty years later Sylvester was invited to be the first professor of mathematics at the new Johns Hopkins University in Baltimore. At the age of 63, he happily returned.

Round about the same time, Playfair, now aged 60, was making his own contribution to links between British and American relationships by marrying the daughter of a prominent Bostonian. His third wife and her family introduced him to the intellectual life in Boston. He particularly enjoyed meeting Longfellow, who told him with glee that an Englishman had called upon him without an introduction, and had apologised by saying: 'Mr. Longfellow, as there are no ruins in this country for a traveller to look at, I have come to see you!'.[31] Playfair was already aware of Boston's literary reputation, but he had not expected the same level of cultural life to be found elsewhere in the States. Chicago, for example:

But what surprised me more than its mushroom quickness of growth, after the great fire, was the state of intellectual development among the citizens. On the morning (it being Sunday) after my arrival I went to the public theatre, to hear a sermon . . . The audience was singularly well dressed, much better than one in the Royal Institution in London. . . . The sermon was remarkable, the subject being 'The Darwinian law of development applied to the coming of the Kingdom of God'. The preacher was thoroughly acquainted with the views of Darwin, Spencer, Huxley, Tyndall and others; but what surprised me most was that the audience seemed also to be. I was so struck with the breadth and eloquence of this sermon that I sent the newspaper containing a full report of it to Darwin, who afterwards expressed his interest in it.[32]

British scientists tended to lump academic posts in the USA together with those in such countries as Canada and Australia. If a good position was not available in Britain, then a chair at one of the universities in these countries was an acceptable alternative. Playfair, himself, was approached by Faraday in the 1840s to see if he would be interested in a professorship of chemistry in Toronto. He began negotiations, but Robert Peel, who was then Prime Minister, stepped in and persuaded him to stay in Britain. A decade later, Tyndall tried for several overseas chairs before finally obtaining an appointment in London. As this implies, competition for chairs abroad was often keen, and the quality of the candidates often high. William Bragg is a good example. He was Third Wrangler at Cambridge in 1884, and, shortly afterwards, went out to Australia to take up a chair at the University of Adelaide. He stayed there for over twenty years, before returning to be professor at Leeds. In 1915, he and his son were awarded the Nobel Prize for physics. Nor were these the only countries offering attractive opportunities. The Japanese decided to implement a crash programme in science and engineering in the 1870s. After a knife-edge vote, it was decided to use English as the main language of instruction, rather than German. As a result, a number of British scientists and engineers went out to Japan as professors. One was William Ayrton, a pioneer of electrical engineering. He later returned to England, where he was Armstrong's colleague at the Central Technical College in London.

In general, Darwin's career matches this picture of overseas interests well. His voyage in the *Beagle* was part of the deliberate policy regarding surveying developed at the Admiralty, while the lack of facilities on board ship was simply characteristic of the period. He was one of the scientists who helped prepare the Admiralty manual of scientific advice for naval vessels. (The interesting point was that he was still being treated as primarily an expert in geology only a few years before the appearance of *Origin of species*.) When it came to languages, he seems to have fallen into the category of those who learnt with difficulty (though he was far from being alone in this). His great proponent in Germany, Ernst Haeckel, visited him in the 1860s, and had to communicate with him in broken English. As the visits by Gray and Haeckel indicate, Darwin received overseas scientists at his home, though his health placed some restriction on the number of visits allowed. For the same reason of ill health, he travelled little after his early years. But his correspondence with overseas countries was extensive. Like other eminent scientists of the period, he received many letters from non-scientists overseas, as well as from colleagues abroad. Rayleigh remembered talking with Darwin about his foreign correspondents: 'When I stayed with C. Darwin in 1870 he told of a letter from an American with "You will excuse my remarking that your own remarkable resemblance to an ape must have unduly influenced your views" '.[33] This might be compared with Haeckel's glowing description:

... tall and venerable ... with the broad shoulders of an Atlas that bore a world of thought; a Jove-like forehead, as we see in Goethe, with a lofty and broad vault, deeply furrowed by the plough of intellectual work. The tender and friendly eyes were overshadowed by the great roof of the prominent brows. The gentle mouth was framed in a long, silvery white beard.[34]

Looking backwards

Т HE early years of the twentieth century, after the death of Queen Victoria, were inevitably a time of reflection. Commentators on the Victorian era often mentioned the rise of science as an essential part of their assessment: 'Of all human activities and developments none are more characteristic of the Victorian Era than those clustering round the word Science'.[1] There was a craze for stories of the future in the latter years of the nineteenth century. One by the American, Edward Bellamy, provides the title for this chapter. Bellamy wrote it from the viewpoint of a narrator in the year 2000 looking backwards to the 1880s. He envisaged a society that had been revolutionised by science and technology. The book became a best-seller (and was adopted as a basic text by the theosophists). Science had clearly come centre stage: does this mean that the efforts of the scientists had finally paid off?

Education and research in science

In terms of education, science was little encouraged in Britain till the last quarter of the nineteenth century. Matthew Arnold, hardly a devotee of science, was sent to the Continent at the end of the 1860s to compare school teaching there and here. Speaking with his school inspector's hat on, he asserted that: 'In nothing do England and the Continent at the present moment more strikingly differ than in the prominence which is now given to the idea of science there, and the neglect in which this idea still lies here'.[2]

Between Arnold's assessment and the end of the century, the figures suggest a considerable growth in the number of people involved one way or another in science education. As one illustration, the expenditure of the Science & Art Department on science education nearly doubled between the end of the 1860s and the mid-1880s, and the number of science teachers over the same period increased from about 750 to well over 2,000. As another illustration at the university level, the number of students taking B.Sc. degrees at London University

THE SCHOOLMASTER OF THE FUTURE.

(And the sooner we get him the better.)

BRITISH WORKMAN. "BOTHER YOUR 'OLOGIES AND 'OMETRIES, LET *ME* TEACH HIM SOMETHING USEFUL!"

A *Punch* cartoon supporting the traditional British belief in the superiority of rule-of-thumb methods over theory. From *Punch* (19 November 1887).
The British Library, PP5270

tripled between 1880 and 1900. And, as an illustration of growth in terms of professional interest, membership of the Chemical Society of London increased by a factor of seven between 1860 and 1900. In the 1880s, Geikie proudly surveyed the scientific scene in the journal, *Nature*:

For one school in which science was taught [twenty years ago] there are a hundred where it is taught now. New colleges have been founded in various centres of industry for special instruction in science. New professorships for the cultivation of different branches of science have been established at some of the older seats of learning. Parliament votes an annual sum of £4,000 for the encouragement of original research. New journals for the illustration of scientific progress have been started. Almost every large publishing firm has organised a series of science class-books.[3]

This sounds like a success story – and so it is – but each point raises a query. For example, the growth in science teaching was real, as was the attempt to build contacts with industry. Yet these efforts were not necessarily reaching the essential audience. Huxley, writing a few years after Geikie, made his own assessment:

The fact remains that the wealthy manufacturer . . . sends his son to a classical school to learn Latin and Greek as a preparation for cloth manufacturing, calico printing, engineering or coal mining . . . After his scholastic career, he enters his father's factory at 20 to 24, absolutely untrained in the chief requirements of the business he is called upon to direct, the complex details of which he has never had an opportunity of mastering.[4]

This disdain among the wealthier ranks of society for a scientific education, was still strong at the end of the century. Not long after Queen Victoria died, one of Lockyer's friends wrote to him: 'the greatest drawback to higher scientific training is not the lack of men to teach or institutions for teaching – though we be here sorely behind other rival nations – but in the pig-headed apathy of manufacturers who won't admit the scientific and trained assistant'.[5] Though the provision of an adequate training in science was a start, recognition of its value took much longer. The same was true of scientific research. At the end of the century, scientists were still bewailing how far British science fell short of activities in Germany. In 1901, Frankland's son, a chemist like his father, compared the amount of chemical research published annually in the two countries. His comment was that the results for Britain and Germany bore, 'somewhat the same relationship to each other as do the homely elevations of the Grampians to the snow-clad peaks of the Andes'.[6]

All scientists agreed on the need for scientific education, but there were dissenting voices on the need for financial support for scientific research from the state. Most were concerned that research in areas of science that had no immediate application would be stifled unless adequate funding was available. But some pointed to the need for care in relying too much on such funds. Airy consistently took this view throughout his life. A characteristic comment (in this case, made in the 1870s) was: 'I think that successful researches have in nearly every instance originated with private persons, or with persons whose positions were so nearly private that the investigators acted under private influence, without incurring the danger attending connection with the State'.[7] Airy's emphasis was on what might be called the 'amateur' strand in British science – the need to be totally independent.

Professionals and specialists

There are many definitions of what the word 'professional' means. Perhaps the simplest one for scientists in the Victorian era would be a man who devoted much of his time to science and scientific research. Such a person might be paid for doing science, as Faraday was, or rely on his own wealth, as Darwin did. Babbage, when he condemned the lack of professionalism in British science in the 1830s, was referring to the need both for a full-time commitment to science and for financial aid to support such a commitment. From the viewpoint of German scientists, Babbage was quite right to define a professional this way. A few years after Babbage's tirade, Liebig noted disdainfully: 'England is not the land of science; there is only a widely dispersed amateurishness'.[8] By the middle of the century, Babbage was still unhappy with the situation. '"Science in England", he wrote, "is not a profession: its cultivators are scarcely recognised even as a class"'.[9] Yet, by the end of the century, the German concept of a professional scientist was widely accepted. Most people working full-time on science in Britain were now paid, often by an educational establishment. This move towards paid professionals as the dominant group was a considerable change over a relatively few decades. It had the unfortunate side-effect that the divide between professionals and amateurs grew, with the latter increasingly seen as second-rate scientists by the professionals.

The professional science community, itself, was becoming increasingly fragmented by the end of the century. Specialisation already existed in the early years of the century, as the appearance of specialised scientific societies then suggests, but the scientific world still had time for polymaths, such as Whewell. Later, the rapid growth of knowledge made mastery of a range of topics difficult, and scientists came increasingly to be classified in terms of the particular specialism in which they worked. (There was an irony here, for Whewell had originally coined the word 'scientist' – which became widely used towards the end of the century – to emphasise the unity of the different branches of science.) A similar narrowing-down affected publication of research. Darwin was one of the last scientists to publish his research in book form. By the end of the century, new research almost invariably appeared in journals, most of which were aimed at specialist audiences. Even the Royal Society's journal, *Philosophical Transactions*, which was intended to cover the interests of all Fellows, bowed to the trend. Towards the end of the 1880s, it was split into two separate journals: Part A aimed at the physical scientists and Part B aimed at the biologists. At mid-century, the Prince Consort had expressed his fear that the unity of science would be undermined by the growth of specialisation. Fifty years later, his fear was beginning to take material form.

Lister was elected President of the Royal Society in the 1890s. One reason was the hope that he would heal, 'the breach that was forming and beginning to widen between the chemists and physicists on the one hand and the biologists, especially the physicians on the other'.[10] Unfortunately, ideas on who could be described as a 'scientist' were by now becoming more clearly defined, and they excluded most physicians. The number of medical men who were members of the Royal Society dropped rapidly at the end of the century. Nor was medicine the only discipline affected. Lubbock's pet subject of anthropology – still dominated by unpaid enthusiasts – was relegated to the fringes, as were a range of other subjects. *Nature* reported at the beginning of the twentieth century that the Royal Society had decided to limit its sphere of activities to the experimental sciences (so excluding such previously acceptable subjects as archaeology, philology, and so on). The rejected subjects banded together to form the British Academy. This split was not reflected in countries on the Continent, where the whole range of disciplines continued to be represented by a single academy.

Just as the emphasis on experimental science led to a division between disciplines, so also it emphasised the division between the new professional scientist and the amateur. In the same year that the *Origin of species* appeared, Huxley argued against using the word 'naturalist', so commonly applied to Darwin:

[It] unfortunately includes a far lower order of men than chemist, physicist, or mathematician. You don't call a man a mathematician because he has spent his life in getting as far as quadratics [i.e. elementary mathematics]; but every fool who can make bad species and worse genera is a 'Naturalist'.[11]

The part-time enthusiast typically concentrated on some area of observational science, particularly on collecting and classifying – the traditional activities of a naturalist in botany, zoology and geology. These aspects no longer captured the interest of the professional, and so the amateur/professional divide grew. The professor of botany at Cambridge had grown up under the old regime. At the end of the 1880s, he bemoaned its passing:

It is rare now to find an Undergraduate or B.A. who knows, or cares to know, one plant from another, or distinguish insects scientifically. I am one of those who consider this to be a sad state of things. I know that much of what is called Botany is admirably taught among us; but it is not what is usually known as Botany outside the Universities, and does not lead to a practical knowledge of even the most common plants.[12]

Women in science

Some of the leading amateurs were women, for observational science, and especially botany, was regarded as an acceptable activity for them. Margaret Gatty, the wife of a Yorkshire clergyman, wrote a widely used text on British seaweeds. An entry in her diary says it all: 'Set off for Filey, Alfred, self, seven children, two nurses and the cook. Arrived safely. Went down to the sands and found seaweeds'.[13] Expertise in foreign languages was also acceptable. The wives of both Sabine and Lockyer translated German and French science books into English. A third form of female scientific activity was to work with one's husband. Huggins' choice of wife seems to have depended, in part, on her interest in helping his astronomical observing. Nor was he alone in this. One of Lockyer's friends declined an invitation to an 'at home' with the explanation: 'it is appointed that I shall be married to the eminent astronomer Miss Klumpke who has resigned her position at the Paris Observatory to be united with me as my wife in the furtherance of astronomical researches at my Observatory'.[14] Yet despite the restrictions, a few women made themselves a name in traditionally

The science writer Mary Somerville in the later years of her life.
The British Library, 10002.b.11

male areas of science. For example, Byron's daughter, Lady Lovelace, was a close friend of Charles Babbage, and wrote extensive notes on the operations of his calculating machine. But the best known figure was Mary Somerville. She was mainly self-taught, but her attempts to produce a unified description of natural phenomena made her name widely known in the scientific community. Her first book, *Mechanism of the heavens*, published in 1832, was used as a course text at Cambridge, though Lyell records Sedgwick's belief that few of the undergraduates would fully understand it. Sedgwick added, 'It is most decidedly the most remarkable work published by any woman since the revival of learning'.[15] Whewell's first use of the word 'scientist' in print was actually in a review of this book. In the mid-1830s, Peel recommended her for a civil list pension, telling her that it was, 'to encourage others to follow the bright example which you have set, and to prove that great scientific attainments are recognised among public claims'.[16]

Mrs. Somerville was a strong supporter of better education for women, and especially that there should be more opportunities for them in science. Not long after her death in the 1870s, a new women's college at Oxford was named after her. By that time, the obstacles to women's participation in science were gradually diminishing. The strength of the opposition was, perhaps, more evident in the learned societies, than in educational circles. Hertha Ayrton, the second wife of Professor Ayrton of the City and Guilds Institute, was, like him, concerned with electrical engineering. She was put up for election to the Royal Society at the turn of the century, but was rejected: the Society's first female Fellow was not actually elected until the mid-twentieth century. (Ayrton was one of those in scientific circles who believed in the 'new woman'. His first wife had been a pioneer female medical student at Edinburgh.) Other societies, too, were debating the question of female membership at this time. For example, the Royal Geographical Society was involved in a furious debate during the 1890s on the admission of women members. (It was not finally resolved in favour of admitting women until just before the First World War.) One opponent outlined the objections:

We contest *in toto* the general capability of women to contribute to scientific geographical knowledge. Their sex and training render them equally unfitted for exploration, and the genus of professional globe-trotters with which America has lately familiarised us is one of the horrors of the latter end of the 19th century.[17]

Hertha Ayrton lecturing to the Society of Electrical Engineers. Note not only the masculine audience, but also the visual aids she is using. From the *Graphic* (1 April 1899). Newspaper Library, Colindale.

Scientists and religion

In a sense all these disputes were internal matters: they concerned who should be accepted into the scientific community. Perhaps more important in the progress of science was the impact that it exerted on the Victorian community at large. An obvious example is the interaction of science and religion. George Lewes, the partner of the novelist George Eliot, tersely summarised this as: 'Religion and Science – the two mightiest antagonists'.[18] This was in the 1870s, when the debate between protagonists on the two sides was at its peak. But the number of vociferous debaters was actually quite limited. Most scientists differed little from other intellectuals of the time in their attitude to religion. They were typically either moderate believers, or moderate doubters. The description of the Church of England as the Tory party at prayer has some resonance here. For example, Rayleigh held family prayers every morning, and Stokes was churchwarden of his local church. Faraday, as a member of a strict sect, was more radically involved in religion. For some years, he was expected to preach on alternate Sundays, and he was once admonished for allowing an invitation from the Queen to interfere with his religious activities. On the liberal side, Hooker remained a member of the Church of England, despite his support for evolution. Contact with clergy was commonplace on this wing, especially where good works were concerned. For example, Lockyer helped in developing Toynbee Hall in the East End of London. This was run by Canon Barnett as a way of introducing university graduates to the problems of the London slums. But it was not people like these who were usually seen as the religious face of science by the public at large. The scientists whose views received the most publicity were those who attacked religion, especially Huxley and Tyndall.

Natural theology – essentially the belief that nature reflected God's activities – was popular in Britain in the early decades of the nineteenth century. By itself, this might not have caused a problem. After all, Paley's text on *Natural theology* was one of Darwin's favourite books as an undergraduate. The problem was that the Bible was interpreted in fairly literal terms as describing the past history of the Earth, and the science of the day was expected to accommodate itself to that understanding. This was particularly difficult for geology, where research was showing that the Earth was much older than a literal reading of *Genesis* would allow. Buckland, as both a clergyman and a

Fig. 4.—*TABLE OF STRATIFIED ROCKS.*

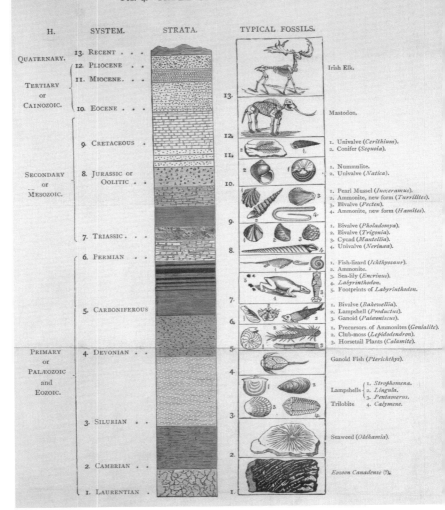

H.	SYSTEM.	STRATA.	TYPICAL FOSSILS.
QUATERNARY.	13. RECENT . . .		
	12. PLIOCENE . .		Irish Elk.
TERTIARY or CAINOZOIC.	11. MIOCENE . . .		
	10. EOCENE . . .		Mastodon.
SECONDARY or MESOZOIC.	9. CRETACEOUS .		1. Univalve (*Cerithium*). 2. Conifer (*Sequoia*).
	8. JURASSIC or OOLITIC . .		1. Nummulite. 2. Univalve (*Natica*).
			1. Pearl Mussel (*Inoceramus*). 2. Ammonite, new form (*Turrilites*). 3. Bivalve (*Pecten*). 4. Ammonite, new form (*Hamites*).
	7. TRIASSIC . . .		1. Bivalve (*Pholadomya*). 2. Bivalve (*Trigonia*). 3. Cycad (*Mantellia*). 4. Univalve (*Nerinæa*).
	6. PERMIAN . . .		1. Fish-lizard (*Ichthyosaur*). 2. Ammonite. 3. Sea-lily (*Encrinus*). 4. Labyrinthodon. 5. Footprints of *Labyrinthodon*.
	5. CARBONIFEROUS		1. Bivalve (*Bakewellia*). 2. Lampshell (*Productus*). 3. Ganoid (*Palæoniscus*).
			1. Precursors of Ammonites (*Goniatite*). 2. Club-moss (*Lepidodendron*). 3. Horsetail Plants (*Calamite*).
PRIMARY or PALÆOZOIC and EOZOIC.	4. DEVONIAN . .		Ganoid Fish (*Pterichthys*).
	3. SILURIAN . .		Lampshells { 1. Strophomena. 2. Lingula. 3. Pentamerus. } Trilobite 4. Calymene.
	2. CAMBRIAN . .		Seaweed (*Oldhamia*).
	1. LAURENTIAN .		Eozoon Canadense (?).

A chart summarising nineteenth-century work on rock strata and fossils. These results offered a major challenge to the ideas of creation circulating in Britain in the early years of the nineteenth century.
The British Library, 7004.a.2

geologist, faced this problem directly, arguing against a literal interpretation:

Minds which have been long accustomed to date the origin of the universe, as well as that of the human race, from an era of about six thousand years ago, receive reluctantly any information, which if true, demands some new modification of their present ideas of cosmogony; and, as in this respect, Geology has shared the fate of other infant sciences, in being for a while considered hostile to revealed religion; so like them, when fully understood, it will be found a potent and consistent auxiliary to it, exalting our conviction of the Power, and Wisdom, and Goodness of the Creator.[19]

Despite this, Buckland was himself later criticised by other scientists for allowing biblical history too large a role in his interpretation of geology. By the latter part of the century, most scientists had swept this type of problem under the carpet, and no longer tried to reconcile their work with events recorded in the Bible. By this time, too, the intention of the scientific radicals had become much wider: not only to deny that natural theology had any meaning, but also to deny that religion needed to be considered by scientists at all. This was the line taken by Huxley, Galton, Tyndall, Spencer and Clifford. Huxley coined the word 'agnostic' to describe the position he and his friends had adopted. For them, the natural world contained all that was knowable. Anything beyond that was uncertain. Huxley propagated his views whenever opportunity offered: he particularly enjoyed slapping down Gladstone's traditional approach. As he wrote to a friend:

From a scientific point of view Gladstone's article was undoubtedly not worth powder and shot. But, on personal grounds, the perusal of it sent me blaspheming about the house with the first healthy expression of wrath known for a couple of years – to my wife's great alarm.[20]

If Huxley was belligerent about religion, Tyndall was even more so. Indeed, his presidential address to the British Association in 1874, widely seen as a major attack on religion, marked the peak of the science *versus* religion controversy. Huxley's later exchanges with Gladstone were seen rather as a kind of sporting contest: the worlds both of science and religion had moved on. From the viewpoint of the clergy, the controversy over evolution, and materialism more generally, was important, but less important than current disputes over the nature of the Bible. Studies under the heading of 'biblical criticism' were typically carried out by clergy, and their disputes, unlike the debate over evolution, involved all the bitterness of a family struggle. In any case, it was recognised that Huxley and his pugnacious friends were not necessarily representative of scientists in general. The view of many scientists was expressed at the end of the 1880s by a senior figure at the British Museum (Natural History):

It is very hard and very unfair that, because Huxley and Tyndall happen to be scientific men of the first order, and happen also to be opposed in some sense to the truths of religion, scientific men generally should be ticketed as though they belonged to the same school of thought.[21]

Scientists and spiritualism

One of Huxley's students at South Kensington was Annie Besant. (He described her as, 'a very well-conducted lady-like person'.[22]) She initially rejected orthodox Christianity even more firmly than Huxley. Later in life, however, she took up theosophy – an attempt to bring together religion, philosophy and science. This shift of interest from the material to the spiritual became commonplace in the latter part of the nineteenth century. Spiritualism – that is, attempts to contact departed spirits – rose rapidly in popularity. The reactions of scientists differed considerably. Huxley and his friends, as might be expected, regarded the whole thing with suspicion. When Huxley was invited to join a committee to investigate spiritualism, he replied:

The only case of 'Spiritualism' I have had the opportunity of examining into myself, was as gross an imposture as ever came under my notice. But supposing the phenomena to be genuine – they do not interest me. If anybody would endow me with the faculty of listening to the chatter of old women and curates in the nearest cathedral town, I should decline the privilege, having better things to do. And if the folk in the spiritual world do not talk more wisely and sensibly than their friends report them to do, I put them in the same category. The only good I can see in the demonstration of the truth of 'Spiritualism' is to furnish an additional argument against suicide.[23]

Huxley was prepared to be amused by spiritualism. Faraday saw it as an offence to religion, as well as to science. After carrying out some experiments on table-turning in the 1850s, he noted:

What a weak, credulous, incredulous, unbelieving, superstitious, bold, frightened, what a ridiculous world ours is, as far as concerns the mind of man. How full of inconsistencies, contradictions, and absurdities it is. I declare that, taking the average of many minds that have recently come before me (and apart from the spirit that God has placed in each) and accepting for a

A Punch comment on Darwin's *Origin of species*. From *Punch* (18 May 1861).
The British Library, PP5270ah

moment that average as a standard, I should far prefer the obedience, affections, and instinct of a dog before it.[24]

Huxley and Faraday were not the only scientists to investigate spiritualism. The most enthusiastic investigator was Crookes, who became firmly convinced that the phenomena were genuine. He believed that a series of crucial experiments into psychic forces he carried out during the 1870s clearly demonstrated their existence. But his attempts to interest the Royal Society and the British Association failed. The members of both were more inclined to sympathise with one current description of Crookes' work as being 'Crookesial experiments into sly-kick forces'. (The phrase reflects how spirit mediums were believed to fake such things as table-turning.) Eventually, Crookes had to give up his investigations when it became apparent that his scientific reputation was suffering as a result. Wallace was another believer in the importance of spiritualism. In 1898, he published a book, *The wonderful century*, which looked back over scientific developments in the nineteenth century. Wallace had no doubts about the progress that had been made, pointing to, 'the chorus of admiration for the marvellous inventions and discoveries of our own age, and especially for those innumerable applications of science which now form part of our daily life, and which remind us every hour of our immense superiority over our comparatively ignorant forefathers'.[25] Yet he believed that there had been failures, as well as successes, and claimed that opposition to psychical research was one of these. To be fair, when the Society for Psychical Research was set up in 1882, a number of Crookes' and Wallace's fellow-scientists did take the trouble to join. (It was open to both believers and sceptics.) Lodge, a believer in spiritualism, joined, but so did Rayleigh and J. J. Thomson, who had no strong opinions either way, and were more representative of scientists in general. As with religion, so with spiritualism, the scientists who made the most noise, either for or against, were not typical of the science community as a whole.

Scientists and vivisection

That is not to say that scientists never took distinctive stands. They did if they saw that important aspects of science were under attack. A good example was the debate about vivisection. During the nineteenth century, the British love of pets rose rapidly, along with opposition to

The superbly moustachioed chemist William Crookes brandishing a glass tube used in his experiments on electrical discharges through gases. From *Vanity Fair* (21 May 1903). Newspaper Library, Colindale

any kind of cruelty to animals. The Society for the Prevention of Cruelty to Animals was set up in the 1820s. Queen Victoria, an enthusiastic supporter throughout her reign, gave it the prefix 'Royal' in 1840. By the 1870s, opposition to experimentation on animals was leading to the demand for rigorous control, or even total abolition of vivisection. Huxley, as a leading physiologist, came under special scrutiny. Needless to say, he reacted vigorously:

The history of all branches of science prove that they must attain a considerable stage of development before they yield practical 'fruits'; and this is eminently true of physiology. Unless the fanaticism of philozoic sentiment overpowers the voice of humanity, and the love of dogs and cats supersedes that of one's neighbour, the progress of experimental physiology and pathology will, indubitably, in course of time, place medicine and hygiene upon a rational basis.[26]

Huxley's position was that vivisection should only be used when it was necessary to advance knowledge, and that the pain inflicted should be minimised as far as possible. As he pointed out, hunting, fishing and shooting – all regarded as admirable activities by most Victorians – were surely more painful to animals than most vivisection, and a good deal less useful. Darwin agreed, and played a significant role behind the scenes in getting the proposed legislation modified. Their position was supported by almost all the scientific community, and not only by the biologists. Kelvin, for example, wrote in support. Lister was the Queen's surgeon (he liked to claim that he was the only man who had ever stuck a knife into her). The Queen, who was strongly anti-vivisection, put pressure on him to support her. He refused to do so. On this occasion, the Queen's camp contained a number of non-scientists whose ideas were normally admired by scientists – Carlyle and Tennyson, for example. In the event, a compromise was reached which allowed scientists to continue their work under carefully controlled conditions, though the keenest anti-vivisectionists (including the Queen) continued to campaign. When the University of Edinburgh celebrated its tercentenary in 1884, Playfair asked the Queen to send a message of welcome. She refused to do so unless she could be assured that no experiments on living animals were permitted at the University. Having delivered her ultimatum, she then left for a visit to Germany. She was pursued there by a series of pleading telegrams, which finally persuaded her to bestow her blessings on the event.

Scientists at the end of the century

Here are three pictures of the interactions between scientists and non-scientists in those key decades of 1860, 1870 and 1880. In the first – evolution and religion – the scientists effectively sparked off the debate. In the second – spiritualism – scientists and non-scientists joined in together. In the third – vivisection – the pressure for the debate came from non-scientists. Yet, in each case, the contemporary discussions confirm that scientists were now recognised, both by themselves and by the rest of the intellectual world, as a distinct group with a distinctive contribution. By the end of the century, scientists had come to be seen as professionals concerned with their own, often esoteric, subject matter. Correspondingly, general debate became more difficult as non-scientists found it harder to grasp new scientific developments. H. G. Wells' science-fiction stories round the turn of the century can, for example, be seen as an attempt to bring scientific possibilities to the attention of the general public. Wells, like Jules Verne before him, was effectively acting as an imaginative intermediary between the scientists and the general public. Indeed, he commented at the end of the century on the extent to which public communication of science by the leading scientists, themselves, was declining: 'Popular science, it is to be feared, is a phrase that conveys a certain flavour of contempt to many a scientific worker'.[27]

Though the emphasis in scientific research was increasingly on group work in a laboratory, the image of science continued as before. Scientific development, it was still felt, depended mainly on the activities of a limited number of outstanding contributors – the belief expressed in Wordsworth's famous description of Newton as, 'a mind for ever voyaging through strange seas of thought, alone'. This view was held by the scientists, themselves, as well as by the general public. Indeed, though Britain lagged far behind Germany in terms of the amount of science produced in the nineteenth century, it nevertheless contributed many of the individual highlights. One reflection of this can be found in the physical sciences, where it has become the habit to call units of measurement after eminent scientists. Such names must receive international agreement, which means that the scientist concerned must have a high international reputation. Among the eminent physicists mentioned in this book, international units have been named after Joule (and another after his tutor, Dalton), Kelvin, Maxwell, Rayleigh, and Stokes. Faraday has had two units named after him.

ARGUMENTUM AD HOMINEM.

"OH, JOSEPH! TEDDY'S JUST BEEN BITTEN BY A STRANGE DOG! DOCTOR SAYS WE'D BETTER TAKE HIM OVER TO PASTEUR *AT ONCE!*"

"BUT, MY LOVE, I'VE JUST WRITTEN AND PUBLISHED A VIOLENT ATTACK UPON M. PASTEUR, ON THE SCORE OF HIS CRUELTY TO RABBITS! AND AT *YOUR INSTIGATION*, TOO!"

"OH, HEAVENS! NEVER MIND THE RABBITS *NOW!* WHAT ARE ALL THE RABBITS IN THE WORLD COMPARED TO *OUR ONLY CHILD!*"

As this cartoon illustrates, current arguments for and against vivisection were already being rehearsed in Victorian times. From *Punch* (20 July 1889).
The British Library, PP5270

This compares well with the record of any other country in the Victorian era.

The downside of this picture of professional eminence was that scientists were increasingly seen as being different from ordinary people. The eccentric scientist – a popular figure well into the second half of the twentieth century – is already there in Wells' *The first men in the Moon*, published in 1901. More importantly, scientists by the end of the century were often seen as arrogant and unconcerned with the implications of their work. Although Huxley, Tyndall, and others had pressed the case for science successfully, the vigour with which they did so often caused unnecessary antagonism. Hooker, reporting to Darwin on one of Huxley's lectures, observed that Huxley, 'offended the clergy twice without cause or warrant'.[28] Darwin was wiser. When he was visited by Edward Aveling, the consort of Marx's daughter, Eleanor, he asked him why he was so aggressive in pressing his socialist ideas. Darwin believed that too much pugnacity in presenting ideas only created resistance.

Darwin's death illustrates how successfully he combined radical scientific ideas with an acceptable public image. People of great eminence were sometimes buried in Westminster Abbey. Newton lay there, for example, as a pre-eminent scientist from an earlier era. In the latter part of the nineteenth century, John Herschel was buried in the Abbey in 1871. This caused no comment. He was followed by Lyell in 1875. This was a little more controversial, but both Herschel and Lyell had remained members of the Church of England throughout their lives. Darwin, notoriously, had not. The proposal that he, too, should be buried in the Abbey was liable to raise more debate. Darwin died in 1882, assuming that he would be buried in his local churchyard. But his friends immediately started urging that his achievements deserved national recognition. Huxley and Galton got to work. They brought in Spottiswoode, who was then President of the Royal Society, Pritchard, who had tutored the Darwin boys, and Lubbock, who, as a Member of Parliament, could lean on his fellow M.P.s, together with a host of others. In the event, the clergy accepted, in some cases with enthusiasm, the appropriateness of Darwin's burial in the Abbey, and there, in April 1882, he was laid to rest. Not every eminent Victorian recommended for burial in the Abbey was accepted. Both Herbert Spencer and Mary Somerville were proposed, but were turned down (thus keeping sociologists and women scientists at arm's length). It is true, of course, that neither had the intellectual eminence of Darwin, Lyell, or Herschel.

When Queen Victoria first came to the throne, science was a recognised part of British cultural life, but not a particularly important part. These burials in the Abbey are one indication among many that, by the latter years of the century, the importance of science in the intellectual life of the nation was widely accepted. Most great Victorian scientists died peacefully at a ripe old age. (Tyndall, typically, was different – he was accidentally poisoned by his wife.) Most were ushered out of this world with appropriate hymns and readings. Darwin had a specially written hymn for his funeral based on a quotation from *Proverbs*: 'Happy is the man that findeth wisdom'. If we consider the profound changes that our group of scientists had helped introduce during the course of the nineteenth century, then the correct reading for the group as a whole must surely be taken from *Genesis*: 'There were giants in the earth in those days'.

References

The full reference is given when a source is cited for the first time in a chapter. Subsequent citations in that chapter to the same source are given in an abbreviated form.

INTRODUCTION

1 B. Webb, *My apprenticeship* (Longmans Green, London, 1926), p.123.
2 Various estimates of the value of nineteenth-century money in terms of present day money can be found. I have used data from the following website: www.ex.ac.uk/~RDavies/arian/current/howmuch.html

CHAPTER 1

1 F. Darwin, *Charles Darwin* (John Murray, London, 1902), p.8.
2 Ibid., p.11.
3 W. Airy, *Autobiography of Sir George Biddell Airy* (Cambridge University Press, Cambridge, 1896), p.19.
4 V. L. Hilts, 'A guide to Francis Galton's "English Men of Science" ', *Transactions of the American Philosophical Society*, vol. 65/5 (1975), p.42.
5 A. Treneer, *The mercurial chemist* (Methuen, London, 1963), p.12.
6 D. W. Forrest, *Francis Galton: the life and work of a Victorian genius* (Paul Elek, London, 1974), p.9.
7 S. P. Thompson, *The life of William Thomson* (Macmillan, London, 1910), vol. i, p.11.
8 V. L. Hilts, 'A guide to Francis Galton's "English Men of Science" ', p.48.
9 H. E. Roscoe, *The life and experiences of Sir Henry Enfield Roscoe* (Macmillan, London, 1906), p.15.
10 J. J. Thomson, *Recollections and reflections* (Macmillan, New York, 1937), p.3.
11 A. J. Meadows, *Communicating science* (Academic Press, San Diego, 1998), p.79.
12 A. J. Meadows and W. H. Brock, 'Topics fit for gentlemen: the problem of science in the public school curriculum', in B. Simon and I. Bradley (eds.), *The Victorian public school* (Gill and Macmillan, Dublin, 1975), p.99.
13 R. J. Strutt, *Life of John William Strutt, Third Baron Rayleigh* (University of Wisconsin Press, Madison, 1968), p.16.

14   Mrs. Lyell, *Life, letters and journals of Sir Charles Lyell* (John Murray, London, 1881), vol. i, pp.14–17.

15   L. Campbell and W. Garnett, *The life of James Clerk Maxwell* (Macmillan, London, 1882), p.27.

16   C. Babbage, *Passages from the life of a philosopher* (Longman, Green, Longman, Roberts, & Green, London, 1864), p.8.

17   Ibid., pp.11–12.

18   J. J. Thomson, *Recollections and reflections*, p.5.

19   A. Treneer, *The mercurial chemist*, p.175.

20   G. A. Hutchison, *Indoor games and recreations* (The Religious Tract Society, London, 1891), p.6.

21   J. J. Thomson, *Recollections and reflections*, p.6.

22   E. Frankland, *Sketches from the life of Edward Frankland* (Spottiswoode, London, 1902), p.4.

23   H. Spencer, *An autobiography* (Williams and Norgate, London, 1904), vol. i, p.86.

24   C. Babbage, *Passages from the life of a philosopher*, p.19.

25   W. Airy, *Autobiography of Sir George Biddell Airy*, p.16.

26   V. L. Hilts, 'A guide to Francis Galton's "English Men of Science" ', p.50.

27   Ibid., p.60.

28   M. J. G. Cattermole and A. F. Wolfe, *Horace Darwin's shop* (Adam Hilger, Bristol, 1987), p.124.

29   H. Spencer, *An autobiography*, vol. i, p.71.

30   E. Frankland, *Sketches from the life of Edward Frankland*, p.17.

CHAPTER 2

1   A. Desmond and J. Moore, *Darwin* (Michael Joseph, London, 1991),   p.26.

2   F. Darwin, *Charles Darwin* (John Murray, London, 1902), p.11.

3   Ibid., p.18.

4   J. J. Thomson, *Recollections and reflections* (Macmillan, New York, 1937), p.61.

5   C. G. Knott, *Life and scientific work of P. G. Tait* (Cambridge University Press, Cambridge, 1911), p.11.

6   J. J. Thomson, *Recollections and reflections*, p.37.

7   M. Goldman, *The demon in the aether* (Paul Harris, Edinburgh, 1983), p.58.

8   A. Macfarlane, *Ten British mathematicians* (John Wiley, New York, 1916), p.140.

9   P. Armstrong, *The English parson-naturalist* (Gracewing, Leominster, 2000), p.119.

10 W. V. Ball, *Reminiscences and letters of Sir Robert Ball* (Cassell, London, 1915), p.32.

11 L. Campbell and W. Garnett, *The life of James Clerk Maxwell* (Macmillan, London, 1882), p.115.

12 J. C. Shairp, P. G. Tait and A. Adams-Reilly, *Life and letters of James David Forbes* (Macmillan, London, 1873), p.37.

13 W. Reid, *Memoirs and correspondence of Lyon Playfair* (Cassell, London, 1899), p.51.

14 D. S. L. Cardwell, *The organisation of science in England* (Heinemann, London, 1972), p.50.

15 M. M. Gordon, *The home life of Sir David Brewster* (Edmonston and Douglas, Edinburgh, 1869), p.55.

16 J. C. Shairp *et al.*, *Life and letters of James David Forbes*, p.54.

17 S. Douglas, *The life of William Whewell* (Kegan Paul, London, 1881), p.51.

18 L. Huxley, *Life and letters of Thomas Henry Huxley* (Macmillan, London, 1900), vol. i, p.20.

19 W. H. Brock, *H. E. Armstrong and the teaching of science, 1880–1930* (Cambridge University Press, London, 1973), p.4.

20 O. Lodge, *Past years* (Hodder and Stoughton, London, 1931), p.70.

21 W. Reid, *Memoirs and correspondence of Lyon Playfair*, p.35.

CHAPTER 3

1 L. Huxley, *Life and letters of Thomas Henry Huxley* (Macmillan, London, 1900), vol. i, p.23.

2 A. R. Wallace, *My life* (Chapman & Hall, London, 1908), p.144.

3 A. Geikie, *Life of Sir Roderick I. Murchison* (John Murray, London, 1875), vol. ii, p.333.

4 W. Reid, *Memoirs and correspondence of Lyon Playfair* (Cassell, London, 1899), p.56.

5 E. Frankland, *Sketches from the life of Edward Frankland* (Spottiswoode, London, 1902), p.127.

6 S. P. Thompson, *The life of William Thomson* (Macmillan, London, 1910), vol. i, p.297.

7 Anon., *History of the Cavendish Laboratory: 1871–1910* (Longmans, Green, London, 1910), p.64.

8 L. Campbell and W. Garnett, *The life and letters of James Clerk Maxwell* (Macmillan, London, 1882), p.389.

9 O. Lodge, *Past years* (Hodder and Stoughton, London, 1931), p.153.

10 L. Huxley, *Life and letters of Thomas Henry Huxley*, vol. i, p.377.

11 J. J. Thomson, *Recollections and reflections* (Macmillan, New York, 1937), p.20.

12 H. E. Roscoe, *The life and experiences of Sir Henry Enfield Roscoe* (Macmillan, London, 1906), p.108.

13 A. J. Meadows, *Science and controversy* (Macmillan, London, 1972), p.134.

14 W. H. Brock, *The Fontana history of chemistry* (Fontana, London, 1992), p.186.

15 Anon., *History of the Cavendish Laboratory*, p.8.

16 C. Babbage, *Passages from the life of a philosopher* (Longman, Green, Longman, Roberts, & Green, London, 1864), p.112.

17 B. Lovell (ed.), *The Royal Institution Library of Science: Astronomy* (Elsevier, Barking, 1970), p.279.

CHAPTER 4

1 H. Spencer, *An autobiography* (Williams and Norgate, London, 1904), vol. ii, p.115.

2 L. Huxley, *Life and letters of Thomas Henry Huxley* (Macmillan, London, 1900), vol. i, p.259.

3 A. Geikie, *Life of Sir Roderick I. Murchison* (John Murray, London, 1875), vol. i, p.196.

4 H. B. Woodward, *The history of the Geological Society of London* (Geological Society, London, 1907), p.10.

5 J. L. E. Dreyer and H. H. Turner, *History of the Royal Astronomical Society 1820–1920* (Royal Astronomical Society, London, 1923), p.10.

6 M. B. Hall, *All scientists now* (Cambridge University Press, London, 1984), p.45.

7 C. Babbage, *Reflections on the decline of science in England* (B.Fellowes, London, 1830), p.43.

8 P. C. Mitchell, *History of the Zoological Society of London* (Zoological Society, London, 1929), p.104.

9 A. J. Meadows, *Science and controversy* (Macmillan, London, 1972), p.99.

10 D. S. L. Cardwell, *James Joule* (Manchester University Press, Manchester, 1989), p.180.

11 A. J. Meadows, *Greenwich Observatory* (Taylor & Francis, London, 1975), vol. ii, p.70.

12 A. Geikie, *Life of Sir Roderick I. Murchison*, p.184.

13 J. Morrell and A. Thackray, *Gentlemen of science* (Clarendon Press, Oxford, 1981), p.425.

14 Ibid., p.117.

15 Anon., *Monthly Notices of the Royal Astronomical Society*, vol. 1 (1828), p.49.

16 W. Spottiswoode, *Proceedings of the Royal Society*, vol. 33 (1881), p.56.

17 W. Ramsay, *Nature*, vol. 53 (1896), p.366.

18 S. Vincent, *Nature*, vol. 55 (1896), p.79.

19 L. Huxley, *Life and letters of Thomas Henry Huxley*, vol. i, p.97.

20 E. Frankland, *Sketches from the life of Edward Frankland* (Spottiswoode, London, 1902), p.128.

21 A. J. Meadows, *Science and controversy*, p.25.

22 Ibid., p.29.

23 F. Darwin, *Charles Darwin* (John Murray, London, 1902), p.98.

24 H. B. Woodward, *The history of the Geological Society of London*, p.63.

25 K. Hentschel, *Mapping the spectrum* (Oxford University Press, Oxford, 2002), p.142.

26 W. V. Ball, *Reminiscences and letters of Sir Robert Ball* (Cassell, London, 1915), p.158.

27 A. G. Bloxham, *Nature*, vol. 50 (1894), p.104.

28 A. J. Meadows, 'Access to the results of scientific research: developments in Victorian Britain', in A. J. Meadows (ed.), *Development of science publishing in Europe* (Elsevier, Amsterdam, 1980), p.52.

29 R. J. Strutt, *Life of John William Strutt, Third Baron Rayleigh* (University of Wisconsin Press, Madison, 1968), p.307.

CHAPTER 5

1 A. Treneer, *The mercurial chemist* (Methuen, London, 1963), p.86.

2 Ibid., p.88.

3 H. Bence Jones, *Life and letters of Faraday* (Longmans, Green, 1870), vol. ii, p.115.

4 O. Lodge, *Past years* (Hodder and Stoughton, London, 1931), p.85.

5 J. J. Thomson, *Recollections and reflections* (Macmillan, New York, 1937), p.47.

6 W. V. Ball, *Reminiscences and letters of Sir Robert Ball* (Cassell, London, 1915), p.34.

7 L. Huxley, *Life and letters of Thomas Henry Huxley* (Macmillan, London, 1900), vol. ii, p.414.

8 A. Geikie, *Life of Sir Roderick I. Murchison* (John Murray, London, 1875), vol. ii, p.247.

9 H. G. Wells, *Experiment in autobiography* (Faber and Faber, London, 1984), vol. i, p.175.

10 A. J. Meadows, *Science and controversy* (Macmillan, London, 1972), p.119.

11 J. V. Eyre, *Henry Edward Armstrong 1848–1937* (Butterworths Scientific Publications, London, 1958), p.98.

12 W. H. Brock, *H. E. Armstrong and the teaching of science 1880–1930* (Cambridge University Press, London, 1973), p.57.

13 Ibid., p.72.

14 L. Huxley, *Life and letters of Thomas Henry Huxley*, vol. i, p.254.

15 Ibid., vol. II, p.253.

16 C. A. Russell, *Science and social change 1700–1900* (Macmillan, London, 1983), p.155.

17 O. Lodge, *Nature*, vol. 40 (1889), p.433.

18 P. Levine, *The amateur and the professional* (Cambridge University Press, Cambridge, 1986), p.64.

19 P. Armstrong, *The English parson-naturalist* (Gracewing, Leominster, 2000), p.141.

20 L. Huxley, *Life and letters of Thomas Henry Huxley*, vol. i, p.424.

21 R. Yeo, *Defining science* (Cambridge University Press, Cambridge, 1993), p.83.

CHAPTER 6

1 C. Babbage, *Passages from the life of a philosopher* (Longman, Green, Longman, Roberts, & Green, London, 1864), p.189.

2 E. Frankland, *Sketches from the life of Edward Frankland* (Spottiswoode, London, 1902), p.125.

3 D. MacHale, *George Boole: his life and work* (Boole Press, Dublin, 1985), p.159.

4 W. V. Ball, *Reminiscences and letters of Sir Robert Ball* (Cassell, London, 1915), p.91.

5 Mrs. Lyell, *Life, letters and journals of Sir Charles Lyell* (John Murray, London, 1881), vol. i, p.374.

6 E. E. Fournier d'Albe, *The life of Sir William Crookes* (T. Fisher Unwin, London, 1923), p.88.

7 H. Bence Jones, *Life and letters of Faraday* (Longmans, Green, 1870), vol. ii, p.234.

8 Ibid., p.186.

9 Mrs. Lyell, *Life, letters and journals of Sir Charles Lyell*, vol. i, p.373.

10 M. Gordon, *The home life of Sir David Brewster* (Edmonston and Douglas, Edinburgh, 1869), p.150.

11 A. G. Duff, *The lifework of Lord Avebury (Sir John Lubbock) 1834–1913* (Watts, London, 1924), p.18.

12 A. Desmond and J. Moore, *Darwin* (Michael Joseph, London, 1991), p.626.

13 L. Huxley, *Life and letters of Thomas Henry Huxley* (Macmillan, London, 1900), vol. ii, p.122.

14 H. E. Roscoe, *The life and experiences of Sir Henry Enfield Roscoe* (Macmillan, London, 1906), p.276.

15 Mrs. Lyell, *Life, letters and journals of Sir Charles Lyell*, vol. ii, p.52.

16 R. J. Strutt, *Life of John William Strutt, Third Baron Rayleigh* (University of Wisconsin Press, Madison, 1968), p.56.

17 H. Spencer, *An autobiography* (Williams and Norgate, London, 1904), vol. ii, p.143.

18 F. Darwin, *Charles Darwin* (John Murray, London, 1902), p.36.

19 A. Desmond and J. Moore, *Darwin*, p.488.

20 A. J. Meadows, *Science and controversy* (Macmillan, London, 1972), p.234.

21 S. Douglas, *The life of William Whewell* (Kegan Paul, London, 1881), p.369.

22 M. Gordon, *The home life of Sir David Brewster*, p.223.

23 C. G. Knott, *Life and scientific work of P. G. Tait* (Cambridge University Press, Cambridge, 1911), p.73.

24 C. Babbage, *Passages from the life of a philosopher*, p.194.

25 A. Treneer, *The mercurial chemist* (Methuen, London, 1963), p.63.

26 A. Macfarlane, *Ten British mathematicians* (John Wiley, New York, 1916), p.39.

27 S. Douglas, *The life of William Whewell*, p.33.

28 A. J. Meadows, *Science and controversy*, p.221.

29 N. A. Rupke, *Richard Owen: Victorian naturalist* (Yale University Press, New Haven, 1994), p.332.

30 M. Roberts (ed.), *The Faber book of comic verse* (Faber and Faber, London, 1974), p.241.

31 C. G. Knott, *Life and scientific work of P.G. Tait*, p.93.

32 A. J. Meadows, *Science and controversy*, p.73.

33 M. Goldman, *The demon in the aether* (Paul Harris, Edinburgh, 1983), p.104.

34 M. Roberts (ed.), *The Faber book of comic verse*, p.200.

35 C. Babbage, *Passages from the life of a philosopher*, p.251.

36 A. Pritchard, *Charles Pritchard* (Seeley, London, 1897), p.147.

37 S. P. Thomson, *The life of William Thomson* (Macmillan, London, 1910), vol. ii, p.612.

38 V. L. Hilts, 'A guide to Francis Galton's "English Men of Science"', *Transactions of the American Philosophical Society*, vol. 65/5 (1975), p.42.

39 Ibid., p.43.

40 A. S. Eve and C. H. Creasey, *Life and work of John Tyndall* (Macmillan, London, 1945), p.113.

41 Ibid., p.75.

42 C. Morgan, *The house of Macmillan 1843–1943* (Macmillan, London, 1943), p.57.

43 L. Huxley, *Life and letters of Thomas Henry Huxley*, vol. i, p.150.

44 A. J. Meadows, *Science and controversy*, p.235.

45 J. A. Secord, *Victorian sensation* (University of Chicago Press, Chicago, 2000), p.426.

CHAPTER 7

1 F. Fleming, *Barrow's boys* (Grant, London, 1998), p.1.

2 A. Friendly, *Beaufort of the Admiralty* (Random House, New York, 1977), p.299.

3 W. B. Turrill, *Joseph Dalton Hooker* (Thomas Nelson, London, 1963), p.15.

4 L. Huxley, *Life and letters of Thomas Henry Huxley* (Macmillan, London, 1900), vol. i, p.47.

5 J. F. W. Herschel (ed.), *A manual of scientific enquiry* (John Murray, London, 1851), p.iii.

6 E. Linklater, *The voyage of the Challenger* (John Murray, London, 1972), p.274.

7 N. A. Rupke, *Richard Owen: Victorian naturalist* (Yale University Press, New Haven, 1994), p.83.

8 A. W. Anderson, *How we got our flowers* (Ernest Benn, London, 1956), p.205.

9 A. J. Smithers, *Honourable conquests* (Leo Cooper, London, 1991), Foreword.

10 H. A. Bruck and M. T. Bruck, *The peripatetic astronomer* (Adam Hilger, Bristol, 1988), p.119.

11 Mrs. Lyell, *Life, letters and journals of Sir Charles Lyell* (John Murray, London, 1881), vol. i, p.189.

12 G. Buttmann, *The shadow of the telescope* (Lutterworth, Guildford, 1974), p.39.

13 S. Douglas, *The life of William Whewell* (Kegan Paul, London, 1881), p.108.

14 M. Gordon, *The home life of Sir David Brewster* (Edmonston and Douglas, Edinburgh, 1869), p.266.

15 W. B. Turrill, *Joseph Dalton Hooker*, p.194.

16 L. Huxley, *Life and letters of Thomas Henry Huxley*, vol. ii, p.360.

17 W. Reid, *Memoirs and correspondence of Lyon Playfair* (Cassell, London, 1899), p.43.

18 A. S. Eve and C. H. Creasey, *Life and work of John Tyndall* (Macmillan, London, 1945), p.25.

19 E. Frankland, *Sketches from the life of Edward Frankland* (Spottiswoode, London, 1902), p.106.

20 O. Lodge, *Past years* (Hodder and Stoughton, London, 1931), p.105.

21 C. Babbage, *Passages from the life of a philosopher* (Longman, Green, Longman, Roberts, & Green, London, 1864), p.201.

22 A. Desmond and J. Moore, *Darwin* (Michael Joseph, London, 1991), p.494.

23 N. Reingold (ed.), *Science in nineteenth-century America* (Macmillan, London, 1966), p.190.

24 W. B. Turrill, *Joseph Dalton Hooker*, p.166.

25 S. Newcomb, *The reminiscences of an astronomer* (Harper, London, 1903), p.273.

26 N. Reingold (ed.), *Science in nineteenth-century America*, p.257.

27 A. J. Meadows, *Science and controversy* (Macmillan, London, 1972), p.37.

28 L. Huxley, *Life and letters of Thomas Henry Huxley*, p.460.

29 R. J. Strutt, *Life of John William Strutt, Third Baron Rayleigh* (University of Wisconsin Press, Madison, 1968), p.147.

30 A. J. Meadows, *Science and controversy*, p.104.

31 W. Reid, *Memoirs and correspondence of Lyon Playfair*, p.240.

32 Ibid., p.241.

33 R. J. Strutt, *Life of John William Strutt, Third Baron Rayleigh*, p.45.

34 A. Desmond and J. Moore, *Darwin*, p.539.

CHAPTER 8

1 C. G. Knott, *Life and scientific work of P. G. Tait* (Cambridge University Press, Cambridge, 1911), p.1.

2 M. Arnold, *Higher schools and universities in Germany* (Macmillan, London, 1892), p.198.

3 A. Geikie, *Nature*, vol. 28 (1883), p.385.

4 G. W. Roderick, *The emergence of a scientific society in England 1800–1965* (Macmillan, London, 1967), p.42.

5 A. J. Meadows, *Science and controversy* (Macmillan, London, 1972), p.265.

6 Ibid., p.78.

7 Ibid., p.111.

8 C. A. Russell, *Science and social change 1700–1900* (Macmillan, London, 1983), p.175.

9 C. Babbage, *The Exposition of 1851* (John Murray, London, 1851), p.189.

10 R. B. Fisher, *Joseph Lister (1827–1912)* (Stein and Day, New York, 1977), p.298.

11 L. Huxley, *Life and letters of Thomas Henry Huxley* (Macmillan, London, 1900), vol. i, p.165.

12 D. E. Allen, *The naturalist in Britain* (Penguin Books, Harmondsworth, 1978), p.185.

13 Ibid., p.131.

14 A. J. Meadows, *Science and controversy*, p.281.

15 Mrs. Lyell, *Life, letters and journals of Sir Charles Lyell* (John Murray, London, 1881), vol. i, p.368.

16 K. A. Neeley, *Mary Somerville* (Cambridge University Press, Cambridge, 2001), p.78.

17 H. R. Mill, *The record of the Royal Geographical Society 1830–1930* (Royal Geographical Society, London, 1930), p.111.

18 F. M. Turner, 'The Victorian conflict between science and religion: a professional dimension', *Isis*, vol. 69 (1978), p.358.

19 W. Buckland, *Geology and mineralogy considered with reference to natural theology* (William Pickering, London, 1836), vol. i, p.8.

20 L. Huxley, *Life and letters of Thomas Henry Huxley*, vol. ii, p.115.

21 O. Chadwick, *The Victorian church* (Adam and Charles Black, London, 1970), part ii, p.4.

22 L. Huxley, *Life and letters of Thomas Henry Huxley*, vol. ii, p.56.

23 Ibid., vol. I, p.420.

24 H. Bence Jones, *Life and letters of Faraday* (Longmans, Green, 1870), vol. ii, p.307.

25 A. R. Wallace, *The wonderful century: its successes and failures* (Swan Sonnenschein, London, 1898), p.1.

26 L. Huxley, *Life and letters of Thomas Henry Huxley*, vol. i, p.434.

27 H. G. Wells, *Nature*, vol. 50 (1894), p.300.

28 A. Desmond and J. Moore, *Darwin* (Michael Joseph, London, 1991), p.560.

Further Reading

Books on individual scientists can be found in the list of references.

Allen, D. E., *The naturalist in Britain: a social history* (Penguin Books, Harmondsworth, 1978).

Barber, L., *The heyday of natural history 1820–1870* (Jonathan Cape, London, 1980).

Brock, W. H., *Science for all: studies in the history of Victorian science and education* (Ashgate, Aldershot, 1996).

Cannon, S. F., *Science in culture: the early Victorian period* (Dawson and Science History Publications, New York, 1978).

Cardwell, D. S. L., *The organisation of science in England* (Heinemann, London, 1972).

Desmond, A., and Moore, J., *Darwin* (Michael Joseph, London, 1991).

French, R. D., *Antivivisection and medical science in Victorian society* (Princeton University Press, Princeton, 1975).

Hall, M. B., *All scientists now: the Royal Society in the nineteenth century* (Cambridge University Press, Cambridge, 1984).

James, F. A. (ed.), *The development of the laboratory* (Macmillan, London, 1989).

—— *'The common purposes of life': science and society at the Royal Institution of Great Britain* (Ashgate, Aldershot, 2002).

Knight, D., *The age of science: the scientific world-view in the nineteenth century* (Basil Blackwell, Oxford, 1986).

MacLeod, R. M., *The 'creed of science' in Victorian England* (Ashgate, Aldershot, 2000).

MacLeod, R., and Collins, P. (eds.), *The parliament of science: the British Association for the Advancement of Science 1831–1981* (Science Reviews, Northwood, 1981).

Morrell, J., and Thackray, A., *Gentlemen of science: early years of the British Association for the Advancement of Science* (Clarendon Press, Oxford, 1981).

Oppenheim, J., *The other world: spiritualism and psychical research in England 1850–1914* (Cambridge University Press, Cambridge, 1985).

Russell, C. A., *Science and social change 1700–1900* (Macmillan, London, 1983).

Topham, J. R., 'Scientific publishing and the reading of science in nineteenth-century Britain: a historiographical survey and guide to sources', *Studies in the History and Philosophy of Science*, vol. 31 (2000), pp.559–612.

Turner, F. M., *Between science and religion: the reaction to scientific naturalism in late Victorian Britain* (Yale University Press, New Haven, 1974).

—— 'The Victorian conflict between science and religion: a professional dimension', *Isis*, vol. 69 (1978), pp.356–76.

Turner, G. L'E. (ed.), *The patronage of science in the nineteenth century* (Noordhoff International Publishing, Leyden, 1976).

Yeo, R., *Defining science* (Cambridge University Press, Cambridge, 1993).

Index

Page numbers in *italics* refer to illustrations